Synthesis and Applications of
Optically Active
Nanomaterials

World Scientific Series in Nanoscience and Nanotechnology*

ISSN: 2301-301X

Series Editor-in-Chief: Frans Spaepen (*Harvard University, USA*)

*For the complete list of volumes in this series, please visit
www.worldscientific.com/series/wssnn

hv

Volume
14

World Scientific Series in
Nanoscience and Nanotechnology

Synthesis and Applications of
Optically Active
Nanomaterials

Feng Bai

Henan University, China

Hongyou Fan

Sandia National Laboratories, USA

W⊘ World Scientific

NEW JERSEY · LONDON · SINGAPORE · BEIJING · SHANGHAI · HONG KONG · TAIPEI · CHENNAI · TOKYO

Published by

World Scientific Publishing Co. Pte. Ltd.

5 Toh Tuck Link, Singapore 596224

USA office: 27 Warren Street, Suite 401-402, Hackensack, NJ 07601

UK office: 57 Shelton Street, Covent Garden, London WC2H 9HE

Library of Congress Cataloging-in-Publication Data

Names: Bai, Feng (Chemist), author. | Fan, Hongyou, author.
Title: Synthesis and applications of optically active nanomaterials /
    Feng Bai, Henan University, China, Hongyou Fan, Sandia National Laboratories, USA.
Description: New Jersey : World Scientific, [2017] | Series: World Scientific series in
    nanoscience and nanotechnology ; volume 14 | Includes bibliographical references.
Identifiers: LCCN 2017024259 | ISBN 9789813222984 (hardcover : alk. paper)
Subjects: LCSH: Nanostructured materials. | Optical materials. | Optoelectronic devices--Materials.
Classification: LCC TA418.9.N35 B35 2017 | DDC 620.1/15--dc23
LC record available at https://lccn.loc.gov/2017024259

British Library Cataloguing-in-Publication Data
A catalogue record for this book is available from the British Library.

Typeset by Stallion Press
Email: enquiries@stallionpress.com

Preface

Synthesis and Applications of Optically Active Nanomaterials comprises excellent reviews in selected topic areas by those who are authorities in their own subfields of nanotechnology with aims to present a comprehensive and coherent distilling of the state-of-the-art experimental results detailed from the otherwise segmented and scattered literature; and to offer critical opinions regarding the challenges, promises, and possible future directions of nanomaterials synthesis and functionalization. The scope of the book covers from nanocrystals and their self-assembly (Chapter 1), synthesis and applications of optically active porphyrin particles (Chapter 2), and synthesis and applications of carbon nanodots (Chapter 3).

We want to thank the contributing authors Wei Wenbo, Li Qi, Feng Bai, Zaicheng Sun, and Hongyou Fan for their time and efforts devoted to the excellent review articles in this book. Ms T. Yugarani from World Scientific Publishing was responsible for much of the coordination necessary to make the publication of this book possible.

Contents

Chapter 1

Self-Assembly of Nanoparticles

Wei Wenbo and Bai Feng

1.1 Introduction

Self-assembly is the process that individual components arrange themselves into a large-scale structure. The ability of nanoscopic materials to self-organize into ordered assembly structures that exhibit unique collective properties has opened up new and exciting opportunities in the field of nanotechnology. Nanometer-sized materials are typically fabricated using either the top-down or the bottom-up strategy. The top-down approach involves crafting of nanoscopic features by controlled removal of materials from larger solids, often through costly lithographic techniques, whereas the bottom-up approach constructs nanoscale structures from smaller components like atoms and molecules. One of the most extensively employed bottom-up methods is the colloidal chemical synthetic route, which is a simple, inexpensive method that can readily be adopted in standard laboratory settings.[1] With this synthetic approach, colloidal nanoparticle of varying sizes, shapes and compositions have been successfully prepared and their properties studied.[2–4]

An intriguing discovery is that colloidal nanoparticles can spontaneously arrange themselves to form larger structures through self-assembly (Figure 1.1).[5,6] This has enabled the construction of amazingly complex structures that are not accessible *via* top-down fabrication techniques. Moreover, the self-assembled structures have been found to exhibit unique collective properties that differ from the properties displayed by the constituent building blocks.[7]

Based on these features and the immense progress in particle assembly methods over the last several years, a widespread interest in this field has manifested

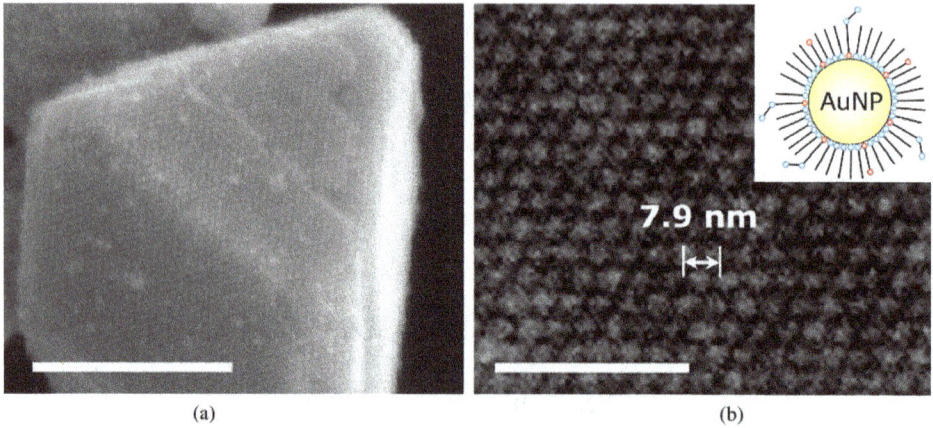

Figure 1.1   Self-assembly of nanoparticles. Individual nanoparticle building blocks prepared by colloidal chemistry techniques ((b), Scale bars: 50 nm) are integrated into ordered structures ((a), Scale bars: 200 nm) by the process of self-assembly without external direction. Adapted with permission from Ref. 5. Copyright 2007 National Academy of Sciences.

itself in a diverse spectrum of research directions and emerging applications, including the following:

(1) *Fundamental Physics.* Spherical colloidal particles typically crystallize in a close-packed face-centred cubic (fcc) lattice which is, beside the difference in size, found also in crystal structures composed of atoms, for example in metals. Building upon this analogy, one can exploit the larger size of colloidal crystals to directly observe fundamental phenomena in solid state physics, such as the formation and propagation of defects and cracks within a material,[8,9] or the transition from a liquid to a glassy or crystalline state.[10] The subject provides further insight into technologically important questions such as packing or jamming and the mutual influence of commensurate or mismatching length scales.[11,12] Diffusion of mesoscopic objects can unravel fundamental concepts of confined mass transport.[13]

(2) *Materials Science.* Hierarchical structuring allows for the realization of robust, yet lightweight materials.[14] Cracking, an often undesired side-effect during colloidal self-assembly, can be specifically put to use to create transparent, conducting electrodes.[15] The colloidal building blocks can be programmed in such a way to add an optical response mechanism to external stimuli, rendering them attractive for future anti-counterfeit applications.[16]

(3) *Energy Migration.* The presence of (periodic) nanostructures can severely influence the way energy or mass is transported through colloidal ensembles. Most prominent examples are the opalescent, structural colors that arise in photonic crystals from the specific interaction of light with the periodic

changes in refractive index.[17] Analogously, mechanical waves can be impeded from traveling through a phononic crystal.[18] Furthermore, the combination of length scales and surface area can contribute to the transport of electrons and analytes in electrode materials or may improve light management in solar cells.[19,20] New endeavors are emerging in the field of thermal transport through colloidally structured materials.[21]

(4) (*Bio*)*Sensing.* Colloidal assembly structures can be rendered sensitive to specific external stimuli (e.g. pH, ionic strength, surface tension), for which in many cases a change of the photonic stopband is used as easily detectable signal.[22] The well-defined nanostructures accessible by colloidal crystals are often exploited to generate plasmonic nanoparticles for use in biosensing.[23]

The beauty of material and structure design with colloids is its simplicity and modular character: well-ordered and complex structures can be created while neither extensive procedures nor expensive equipment are required. The range of structures that can be assembled from colloidal building blocks is vast, which to a large extent is due to the broad definition of what a colloidal particle, the constituting building block of such matter, actually is.

1.2 Building Block in Nanoparticle Self-assembly

Colloidal particles are available from many different materials, like clays, minerals, organic compounds (as in pigments), polymers, ceramics, semiconductors, and metals.[24,25] Colloidal particles have long been used in a broad range of industrial applications such as inks, food, coatings, cosmetics, and rheological fluids. This present chapter includes colloidal particles with size ranging from a few nanometers to a few hundreds nanometers, which may display size-tunable properties due to electron confinement and have come to be known as quantum dots, or nanoparticles. The large particles with dimensions more than 1 μm or small particles with dimensions less than 3 nm (clusters) are not covered here, as additional material parameters such as the nanoparticle shape, the surface functionality of individual facets, as well as additional magnetic, electronic, and dipolar forces, exceed by far the scope and limits of this book. For these assembly structures, we refer the interested reader to dedicated review articles.[26,27]

In order to produce well-ordered assembly structures, it is of key importance that the nanoscale building blocks are uniform in size and shape (i.e. monodisperse). Significant advances in colloidal nanoparticle synthesis coupled with size-sorting processing techniques (e.g. size-selective precipitation) have enabled the preparation of high-quality nanoparticles with a size distribution below 5%.[28] Colloidal nanoparticles with well-controlled size can now be routinely prepared for a vast array of materials, which include metals, metal oxides, and metal chalcogenide semiconductors.[29,30] Oftentimes, the nanoparticles that are synthesized are in the form of spheres (or dots). Owing to the simplicity of their shape and their well-developed synthetic protocols, spherical nanoparticles are the most frequently used elementary components in nanoscale self-assembly.[31] These nanoscopic units usually self-assemble into a compact structure through either the cubic (ccp) or the hexagonal (hcp) close-packing arrangement of spheres.[32]

Tailored surface modification for specific binding sites at defined regions of the colloidal particle provide a higher level of control in directed self-assembly.[33–36] This interesting field is becoming increasingly active with advances being made in the synthesis of highly defined "patchy" colloids with heterogeneous surfaces composed of multiple materials.[37] Analogies drawn from such spatially oriented binding sites with respect to atomic orbitals may provide an efficient access to more versatile crystal structures, much like the richness of crystal structures found in atomic crystals.

1.3 Driven Forces in Self-assembly

We discuss efforts to create next generation materials *via* bottom-up organization of nanoparticles with preprogrammed functionality and self-assembly instructions. This process is often driven by both interparticle interactions and the influence of the assembly environment.

The balance of different forces that act on colloidal particles during their organization in an assembly process can be very complex.[38] It is thus of essential importance to have a general understanding of this balance for obtaining the targeted structural hierarchy with desired complexity. The quantification of the various forces involved in different assembly processes is an important and active research field, yet it is very difficult to generalize.[39] As this chapter is intended to provide an overview on the vast field of colloidal assembly structures, we restrict the discussion to a qualitative description of the most relevant forces necessary to understand colloidal self-assembly processes. Generally, a subtle balance between attractive and repulsive interactions is required to allow probing of local thermodynamic minima by the particles and to overcome activation barriers by the energy provided from the environment to assemble into an ordered structure.

The forces involved in colloidal assembly processes can be divided into three main classes (Figure 1.2): (1) repulsive forces resulting from interparticle interactions between two and more particles (or with extended interfaces) that prevent a colloidal dispersion from spontaneous aggregation and flocculation, (2) attractive interparticle forces that can counteract repulsive forces and may yield specific assembly structures as a result of this (im)balance, and (3) external forces acting on individual particles or particle ensembles.

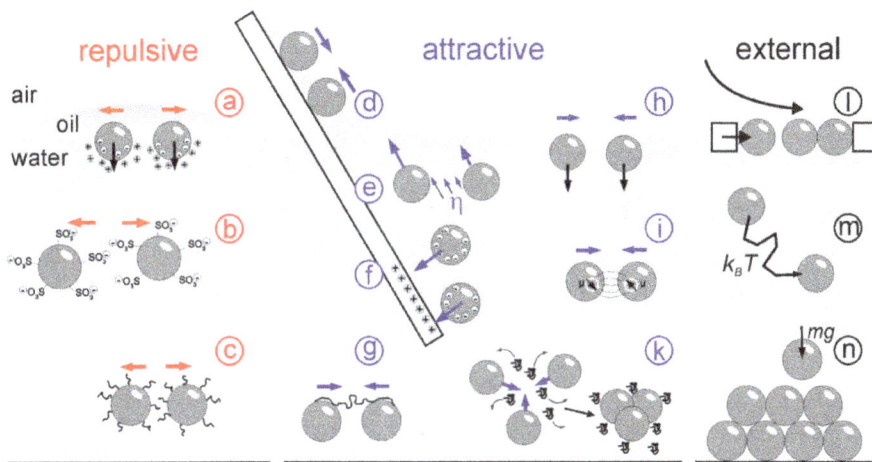

Figure 1.2 Schematic representation of various forces and interaction motifs for nanoparticles occurring in three dimensions and at interfaces. Repulsive interparticle interactions (red): (a) dipolar repulsion by partial ionic dissociation at interfaces, (b) Coulomb repulsion, and (c) steric repulsion. Attractive interactions (blue): (d) immersion capillary forces, (e) hydrodynamic coupling/drag forces, (f) Coulomb attraction to oppositely charged surfaces, (g) bridging attraction/flocculation, (h) flotation capillary forces, (i) van der Waals (vdW) attraction, and (k) depletion attraction. External forces (black): (l) barrier compression or forced convection (effectively resulting in particle aggregation), (m) Brownian motion, and (n) gravitational sedimentation. Adapted with permission from Ref. 38. Copyright 2015 Royal Society of Chemistry.

Details of these forces have been described in previous publications and are reviewed here only in brief to give a qualitative flavor of the large variety of different parameters that have to be taken into account and that can be tuned experimentally to yield the desired structures and order. For further reading reference is given to authoritative overviews in the literature.[40,41]

1.3.1 *Dipolar Interactions*

At the air–water interface, ionic groups at the particle surface can only dissociate inside the water phase, which results in an asymmetric charge distribution around the particle.[42] The resulting dipoles on all particles are oriented parallel to each other and lead to a repulsive force if the ionic surface groups are isotropically distributed around the particle.

A similar scenario is found at the water–oil interface, but partial wetting of the particle surface with water droplets in the oil phase may lead to partial ion dissociation in the oil phase, resulting in stronger particle–particle repulsion by an additional electrostatic component. Furthermore, the presence of an oil–water interface instead of an air–water interface may reduce the counteracting attractive forces given by flotation capillary attraction.[43]

1.3.2 *Coulomb Interactions*

They are governed by the charges distributed over the colloidal object and are repulsive between like-charged particles. Charge densities and spatial distribution functions are expressed by the Poisson–Boltzmann formalism and are approximated by the linearization of Debye. The electrostatic repulsion scales with e^{-kx}, where k represents the inverse Debye length. Electrostatic repulsion can be of long range order; however, due to the self-dissociation of water, it is limited to a maximum of 680 nm. Electrostatic repulsion is one of the major contributors to colloidal stabilization and is summarized in the DLVO theory. The magnitude of electrostatic repulsion can often be tuned by the pH of the dispersion medium or by the addition of salt, which changes the ionic strength of the dispersion medium and thus the Debye length.

1.3.3 *van der Waals (vdW) Forces*

van der Waals (vdW) forces are generally attractive multibody dipolar interactions. Their magnitude and range depend on the size and shape of the colloidal object. This behavior is captured by the Hamaker formalism. Colloidal vdW forces act over much larger distances (force $\propto (1/d^2)$, for two spheres with $R \gg d$) compared to intermolecular vdW interactions. An interesting feature, which can be inferred from the Hamaker formalism, is the fact that vdW forces can be effectively screened by refractive index matching of the dispersion medium to the colloidal objects.[44] This fact can be experimentally exploited to tune the particle–particle interaction.

1.3.4 *Steric Stabilization*

In the case of colloidal particles with a solvated polymer corona, colloidal stability is attained by steric stabilization. The aggregation of colloidal particles is prevented by the evolution of an osmotic pressure in the region of polymer chain overlap (therefore local polymer concentration increase), as well as by mechanical spring forces, which result from an unfavorable decrease in entropy due to compression and restriction of the solvated polymer chains.[45] The range of this interaction force is given by the spatial extension of the stabilizing layer. The use of theta-solvent conditions for the corona polymer eliminates the evolution of an osmotic pressure in the overlap region and thereby may be exploited to reduce the stabilizing effect.

The repulsive forces between particles in liquid media are the major contributors in the stabilization of colloidal dispersions. They are crucial to guarantee a long shelf life and processability.

1.3.5 *Capillary Force*

When a colloidal particle comes into contact with an interface between two liquid media it can be energetically stabilized or "trapped" there with a degree of stabilization up to $10^7 \, k_B T$ (at room temperature) for a particle diameter of 1 μm and equal wetting by the two liquid phases. This interfacial stabilization was explained by Pieranski as the change of the interfacial energy between the two mobile phases in the presence of the particle and the increased contact area of the particle with both phases.[42] This effect is the basis for Ramsden and Pickering emulsions, where dispersions of water and oil are stabilized by solid particles located at the water–oil interface. It also plays an important role in the stabilization of food emulsions, foams, as well as in the industrial flotation process.[46]

The presence of a liquid or solid interface strongly affects the mode of interaction between colloidal particles. Confinement of the particles at an interface usually leads to slower particle dynamics. It is associated with a reduction of the free translational motion in the bulk to a planar random walk at the plane of the interface. Specific interfacial interactions of particles have been discussed in the literature and are only briefly sketched in the following.[47,48]

When the particles deform the liquid interface, the surface tension of the liquid causes an effective interaction between the particles. If the deformation between two particles is in the same direction (which means the particles are wetted in the same way) the interaction is attractive and the resulting capillary forces will pull the particles together. For floating particles, this effect takes place if their diameter and density difference to the liquid medium is large enough to cause an interfacial deformation by gravitational and buoyancy forces (typically for particle dimensions larger than 1 μm). The extent of interfacial deformation further depends on the wettability of the colloidal object with the liquid medium and the particle shape. The resulting interaction is known as the flotation capillary force.[49] For very small non-isotropic particles, this flotation capillary force can be significant, due to the meniscus deformation by other mechanisms such as electrostatic stresses.[50]

When particles are trapped at a solid interface in a liquid film with a thickness below the particle diameter, a deformation of the liquid surface can take place even for particles much smaller than 1 μm. The resulting immersion capillary forces drag the particles together and are often encountered in drying latex dispersions. Hydrodynamic coupling can lead to deviations from the random Brownian motion by the presence of a convective flow of the liquid medium or from viscous coupling between the particles.[49]

1.3.6 *Depletion Forces*

The presence of much smaller objects, such as individual dissolved polymer chains or smaller nanoparticles, can lead to depletion forces, which drive aggregation of the larger particles. Thereby, the overall system reduces its total energy by an increase in the degrees of translational freedom and entropy.[51] The combination of surface roughness with depletion interaction can lead to aggregation discrimination between rough and smooth particles, with the smooth colloids showing preferential depletion attraction.[52]

1.4 Strategy for Self-assembly

The formation of organized structures can take place on a substrate surface, at an interface, or even in the bulk solution, and the choice of an assembly technique is important in achieving the desired structure having properties that can be potentially useful for functional device applications. Ordered arrays of nanoparticles may be prepared by evaporation or destabilization of a nanoparticle solution.[53]

Evaporation-based assembly typically leads to superlattice thin films and takes place at the late stages of solvent drying when particles find themselves in crowded solution. Destabilization-based assembly exploits attractive interactions between nanoparticles when solvent intermingling in nanoparticle capping layers becomes less favorable than overlap of ligands between neighboring nanoparticles, promoting gradual clustering of nanoparticles in solution. A more detailed discussion on each of these techniques is presented in the following subsections.

1.4.1 *Evaporation-mediated Assembly*

One of the most widespread and simplest methods for colloidal crystal fabrication is based on convective particle aggregation. The interplay of convective drag force, electrostatic repulsion, and capillary attraction caused by the evaporation of solvent at the liquid meniscus in contact with the solid substrate leads to highly ordered structures.

The formation of ordered self-assembled nanostructures was first observed in the 1980s using transmission electron microscopy (TEM) when a dispersion of $Fe_{1-x}C_x$ particles in benzine (i.e. petroleum ether) was deposited onto a carbon-coated copper TEM grid and subsequently dried in air. The $Fe_{1-x}Cx$ particles were oxidized to iron oxide in air and were found to form three-dimensional (3D) close-packed structures of monodisperse spherical particles on the amorphous carbon film upon drying. The solvent evaporation rate is believed to be critical to the

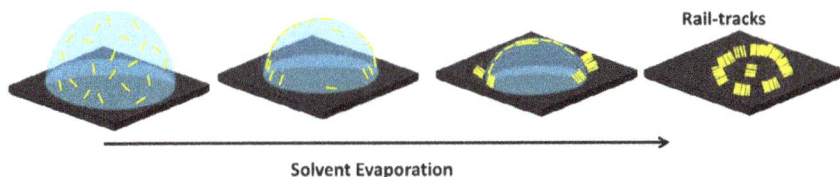

Figure 1.3 Schematic showing the process of solvent evaporation induced assembly of nanorods. Adapted from Ref. 56 with permission from The Royal Society of Chemistry.

formation of these well-ordered structures. In the succeeding years, this simple assembly approach, which was conveniently termed as evaporation-mediated or drying-assisted assembly, has been widely used in the self-assembly of spherical as well as anisotropic nanoparticles on solid substrates.[54,55]

A schematic representation of evaporation-mediated self-assembly of nanorods on a substrate that is based on the method of drop-casting (Figure 1.3).[56] In general, when a drop of nanoparticle dispersion is deposited onto a clean, flat substrate (e.g. silicon, silicon nitride) and the solvent is subsequently evaporated in a controlled manner, the relatively weak attractive forces (e.g. vdW, dipole–dipole interactions) between the dispersed nanoparticles become apparent as the volume of the droplet is reduced, driving nanoparticles to self-organize.[57] Furthermore, forces such as electrostatic repulsive forces, hydrophobic interactions (associated with capping ligands), capillary forces, and entropic depletion interactions can each play a role and mediate the assembly formation.[58]

Most predominantly, direct assembly methods take advantage of solvent evaporation to control the deposition of colloids. Such techniques are usually referred to as convective assemblies and are based on the formation of a very thin liquid film in the meniscus region of at the three-point contact line. The dominating forces governing the crystallization mechanism are immersion capillary forces that push the particles together once the height of the liquid film falls below the colloid diameter.[49]

A convective flow is consequently created as water evaporates in the formed monolayer. This flow opposes random Brownian motion and continuously drags particles from the bulk dispersion to the monolayer nucleus and causes its continuous growth.

Evaporation-induced deposition methods that rely on capillary forces require low particle–surface interactions so that particles can freely diffuse across the substrate, seeking their lowest energy configuration. This is often achieved *via* electrostatic repulsion by chemically modifying the colloid and substrate surface with negatively charged functionalities (such as carboxylate or sulfate moieties). These are repelled by negatively charged substrate surfaces, present for example on glass. In addition, the ordering process is dramatically influenced by the surface energy and wetting properties of the particle and substrate surfaces: to allow for the assembly process to take place, a very hydrophilic substrate with low water contact angles (i.e. below 20°) is crucial.[59] A higher contact angle prevents the formation of monolayers on flat substrates as the water film dewets before the height of the water layer becomes small enough for immersion capillary forces to act effectively.[60]

Evaporation-induced methods can be performed with different levels of sophistication. In the experimentally simplest case, a droplet of water is left to evaporate on a hydrophilic substrate (drop casting). Pioneered by Nagayama in the 1990s, this method has been valuable to study the fundamental mechanisms of the crystallization process that is driven by immersion capillary forces. Although ordered patches of monolayers (~mm^2 dimensions) will form upon evaporation of solvent, a homogeneous coverage of large areas with well-ordered monolayers is impeded by the coffee-stain effect, leading to multilayers at the drying front and sub-monolayer in the central part of the droplet. The circumvention of coffee-stain effects requires more elaborate process designs, including the addition of surfactants[61] or polymers[62] to the deposition solution or the use of anisotropically shaped particles.[63]

In 1996, a vertical deposition method was introduced to form homogeneous, well-ordered monolayers over larger areas. The assembly mechanism is conceptually similar to the drop casting method. However, control over the deposition is gained by slowly removing the substrate from a colloidal dispersion. This enables the formation of a straight meniscus line at which the assembly process takes place. To achieve homogeneous growth of the monolayer, the substrate is slowly moved in the opposite direction to the growth front and a convective flow of particles toward the growth front induces crystallization (convective assembly). Both, evaporation rate and withdrawal speed, have to be matched carefully. Though the technique leads to assembly structures of high order, the uniform coverage over large areas may be hampered as precise control over evaporation conditions can be difficult to achieve. Figure shows typical heterogeneities (voids and multilayers) that are related to the difficulties of exactly controlling the process conditions. The addition of surfactants (such as sodium dodecyl sulfate, SDS) beyond its critical micelle concentration to the particle dispersion has shown to result in a more rapid and robust colloidal monolayer formation, though this has been only demonstrated on a gold surface.[64]

Another variation of the process is the horizontal deposition technique.[60,65] A receding meniscus is formed by a confining glass slide mounted on top of a translational stage. The particle monolayer is formed during horizontal displacement of the target substrate. The advantage of the horizontal technique is the possibility to directly observe the deposition process by optical microscopy. The solvent evaporation rate, which is controllable by the substrate temperature, the target substrate velocity, and the suspension particle concentration needs to be mutually matched for homogeneous monolayer growth. If carefully controlled, the size of the colloidal crystal (i.e. the number of layers) can be adjusted for a single monolayer and a double layer.[59]

In comparison with isotropic spherical nanoparticles, 1D nanoparticles are more challenging to self-assemble due to their inherent anisotropic structure. In assembling colloidal 1D nanoparticles, a very slow evaporation process that takes several hours to complete is usually adopted as this has been found to be beneficial in increasing the packing order and size of the assembled structure. Figure shows a schematic illustration of a more tightly controlled evaporation setup that has been used in the self-assembly of colloidal nanorods on a wide range of substrates. The evaporation chamber is supplied with a controlled flow of dry nitrogen to enable control of the drying rate. To ensure a constant solvent evaporation rate, the nanoparticle dispersion is maintained at a particular temperature, which is typically between room temperature and 60°C depending on the chosen solvent's boiling point and the desired evaporation rate. Temperatures above 60°C have been reported to produce disordered structures as the evaporation is too fast for an organized assembly to occur.[66]

The nanoparticle aggregation and ordering mechanisms have been explained using thermodynamics[67] and coarse-grained lattice-gas models,[68] where the formation of the final ordered structure is thought to be influenced by several factors, including temperature, nanoparticle concentration, nature of solvent, and nanoparticle size, among others. Proper control of these critical parameters is therefore necessary for a successful self-assembly. The effects of nanoparticle concentration and solvent nature on the evaporation-mediated self-assembly of colloidal nanorods of cadmium chalcogenide semiconductors (CdS and CdSe) have been investigated by Ryan and co-workers.[56] When the nanorod concentration is very low, random deposition of the nanorods on the substrate was observed. In this case, the nanorods are very far apart that the attractive forces between them are too weak to trigger self-assembly. A concentration that is too high, however, has only led to short-range ordering, suggesting that the rods are too close that the repulsive forces between them become significant. Thus, for a highly ordered assembly, there exists an optimal concentration window such that the interrod distances are small enough that the attractive forces between the nanorods far outweigh the repulsive forces.

In choosing an appropriate solvent, properties such as volatility, polarity and dielectric permittivity should be taken into consideration. Because the drying rate is largely dependent on the volatility of the solvent, the use of a highly volatile solvent (i.e. having low boiling point) is not recommended as this would lead to evaporation that is too fast. The choice of solvent is also restricted by its ability to effectively disperse the nanorods as poorly solvated nanorods have a high tendency to randomly agglomerate. A non-polar solvent (e.g. toluene, cyclohexane) is well-suited for colloidal nanorods that are surface-passivated with organic surfactants/ligands having large hydrophobic moieties (e.g. trioctylphosphine, oleic acid). Lastly, a solvent with a dielectric constant (ε) lower than 4 is desirable as the strength of the electrostatic forces that facilitates the assembly would be considerably screened in a solvent with higher permittivity. For this reason, chloroform ($\varepsilon = 5$), although non-polar, is not a good solvent for evaporation-mediated assembly of hydrophobically coated nanorods.

Other key parameters that have been shown to affect nanorod self-assembly are the concentration and nature of the stabilizing surfactant (capping ligand), the nanorod aspect ratio, and the interaction energy between the nanorods and the substrate.[69,66]

1.4.2 *Destabilization-mediated Assembly*

Destabilizing solutions of nanoparticles produces close-packed 3D superlattices like platelets, polyhedra, or spheres. In this case, gradual onset of attractive interactions between dispersed nanoparticles induces slow clustering of particles in the solution bulk. For example, Rupich *et al.* found that transferring a layer of non-solvent above a solution of nanocrystals leads to the self-assembly of nanocrystals into flat platelets or multiply twinned polyhedra depending on particle size, as shown in Figure 1.4.[70] This general approach has also been used to assemble nanorod[71] and nanoplatelet[72] superlattices.

Along these lines, functionalizing nanoparticles surfaces with light-triggered molecular switches results in colloid destabilization and similar polyhedral superlattices.[5] Another destabilization-based technique for achieving 3D superlattices involves inducing solvophobic interactions by disrupting a surfactant bilayer.[73] In this approach, dodecyltrimethylammonium bromide (DTAB) surfactant is used to hydrophilize organic-capped nanocrystals, forming a bilayer held together by van der Waals forces between aliphatic chains. Subsequent exposure to polymer-containing ethylene glycol solution at 80°C decomposes the bilayer and leads to the formation of round superlattices with fcc internal packing structure. This approach has also been used to make spherical and needle-shaped superlattices of nanorods.[74]

1.4.3 *Depletion-force-driven Assembly*

While destabilization-based assembly of nanoparticles has typically been carried out by introducing attractive interparticle interactions *via* reduction of solvent

Figure 1.4 Destabilization-based assembly of nanoparticle. (a) Schematic of self-assembly by slow diffusion of the non-solvent into the dispersed nanocrystal colloid. (b) SEM overview of platelet-shaped superlattices formed from destabilization of a toluene solution of 3 nm PbS nanocrystals. (c) and (d) SEM zoom of individual platelet superlattices. (e) SEM overview of multiply twinned superlattices with icosahedral or pentagonal (5-fold) symmetry formed from destabilization of toluene solution of 8 nm PbS nanocrystals. (f) and (g) SEM zoom of individual polyhedral superlattices. Adapted from Ref. 70. Copyright 2010 American Chemical Society.

affinity for surface ligands, entropic destabilization is another means to direct superlattice formation. The depletion effect is achieved by adding small, non-interacting co-solutes (depletants) to the solution, which cause an attractive depletion force between large colloidal nanoparticles. This force arises because the depletant center of mass cannot access the solution within one depletant radius around the surface of an isolated nanoparticle (Figure 1.5).[40]

Overlap of such exclusion zones increases the volume of solution available to the depletants, lowering the free energy of the system by increasing the depletant translational entropy. The result is an effective osmotic pressure of the solution inducing precipitation of the nanoparticles even in good solvent for the nanoparticle surface ligands. For example, phosphonate-capped CdSe/CdS nanorods have been assembled into superlattice sheets by addition of oleic acid (0.5–1.0 M) to an initially clear nanoparticle solution in toluene.[51]

According to the Asakura–Oosawa model,[75] the potential energy arising from depletion of micelles between two particles at contact scales as the product of the excluded volume and concentration of micelles in solution. Geometrically

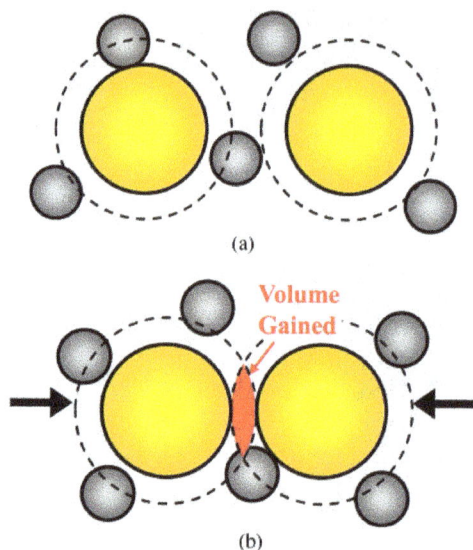

Figure 1.5 Depletion interaction. (a) Illustration of gold nanoparticles immersed in a solution of smaller spherical macromolecules (gray spheres). The dashed lines illustrate the excluded volume regions into which the center of the solute particles cannot enter. (b) As two particles approach, overlap of the excluded volume shells increases the volume available to the solute, decreasing the free energy of the system. Consequently, there is a mean force (black arrows) acting to push the particles together. Adapted with permission from Ref. 40. Copyright 2009 John Wiley and Sons.

speaking, the excluded volume is greatest for large interparticle contact area, i.e. contact between low-curvature surfaces. As a result, the addition of depletants is an effective approach to separate size and shape impurities from a crude nanoparticle synthesis mixture.[76] Depletion is also important for the self-assembly of binary nanoparticle mixtures. The addition of depletants stabilizes superlattices that maximize contact area and/or the number of contacts between larger species.

The manipulation of directional forces *via* surfactant micelle induced depletion interaction aligns nanoparticle facets and leads to additional geometric control and tunability.[77,78] It is also a pathway towards avoiding unwanted mechanical deformation. Depending on the thickness of PVP ligands, Pd cubes assemble into a superlattice that continuously transforms between a simple cubic phase and a rhombohedral phase.[79] An effective rounding or truncation of cube edges and vertices causes a shear of the superlattice, as also found with platinum nanocubes,[80] increasing the effective packing density.[81,82] Adding poly(ethylene oxide) (PEO), depletants suppresses this shear, as demonstrated for micrometer-sized hollow silica cubes grown around hematite templates.[83] Simulations suggest that leveraging the distribution of interparticle voids with polymeric additives can favor either of the two polymorphs, distinguishing the close-packed superlattices fcc and hcp or competing open superlattices.[84,85]

Finally, while the use of depletion is a purely physical effect that is based on entropy maximization, adding other additives allows controlling nanoparticle assembly chemically. For details and ideas we refer to a recent review.[86]

1.4.4 *Gravitational Sedimentation*

A less common approach to assembly nanoparticle superlattices exploits gravitational sedimentation. Since gravity biases thermal motion of nanoparticles with diameter approaching 1 μm, or nanoparticle core materials comprised of high-density metals, crowding-induced self-assembly can occur *via* sedimentation of nanoparticles in the bottom of solvent. The propensity for particles to accumulate in the bottom of a solution under the influence of gravity can be evaluated by comparing the relative size of thermal energy k_BT and the gravitational potential energy mgd required to raise a particle of mass m by its own diameter d in Earth's gravity g. The ratio of k_BT to mgd scales as d^{-4} and for 10 nm nanoparticles, for example, is approximately 10^6, while for micrometer-sized particles it is below 1. Accordingly, even in the presence of repulsive interparticle interactions, the largest nanoobjects (100–1000 nm) are expected to sediment under the force of gravity in all but the densest of liquids.[87] Particles in this size regime exceed the 100 nm boundary set by the community (according to the ISO/TS 80004-2:2015 standard) to distinguish nanoparticles from coarser particulate matter. In fact, most of the nanoparticles discussed in this book fall comfortably below this limit and exhibit assembly behavior not significantly influenced by gravitational forces. Such emphasis primarily reflects that of the self-assembly literature, a result of the availability of synthetic protocols for producing uniform particles at the lower end of the nanoscale, fundamental interest in quantum size effects, and the ease with which smaller nanoparticles diffuse, explore configurational space, and ultimately adopt equilibrated structures than their larger counterparts.

1.5 Effect of Environment in Self-assembly

Akin to the art of protein crystallization, nanoparticle self-assembly is sensitive to several factors beyond the quality of the starting material. For instance, the choices of solvent, temperature, and substrate play a role in the ordering of nanoparticle superlattices. Undesired flocculation of particles in solution before triggering assembly by evaporation or destabilization can suppress ordering. As such, use of a good solvent for aliphatic capping ligands (e.g. hydrocarbon liquids such as hexane, octane, or toluene; chlorinated hydrocarbons such as chloroform, tetra-chloroethylene, chlorobenzene) promote dispersal of the colloid and are good starting point for assembly experiments. Gentle heating of the assembly solution facilitates ordering of nanoparticles in superlattices. Because nanoparticles experience a thermodynamic drive to eliminate surface area if provided sufficient thermal energy to coalesce, thermal decomposition of the material presents a practical upper limit to assembly temperature.[88] In addition, the solvent vapor pressure is an important parameter for evaporative self-assembly experiments. Because the ordering process requires particles to diffuse through solution and sample at various positions, use of volatile solvents may condense particles too rapidly to allow for self-assembly. Furthermore, the choice of support (i.e. solid or liquid subphase) influences the assembly outcome, setting dimensions and orientation of nanoparticle superlattices.

1.5.1 *Patterned Substrate*

The preceding treatment of assembly and interfaces discussed systems with flat surfaces. In this case, surface tension favors a smooth interface between immiscible liquids, and atomic-scale roughness of solid supports is small compared to the diameter of typical nanoparticles. Evaporating a nanoparticle solution over a surface with texture size on the order of the particle diameter presents another strategy for structural control. For example, capillary forces may be used to trap solvated nanoparticles in the recessed regions of lithographically patterned templates.[89–91] Such patterns control the position and number of nanoparticles within a cluster[92] and selectively guide formation of contacts between nanoparticles of different shapes.[93] In addition, sequential deposition of nanoparticle layers (nanoparticle epitaxy[94]) enables templated assembly without top-down lithographic pattern generation. In this approach, a base layer of nanoparticles immobilized by plasma treatment serves to guide the assembly of a second layer. Adlayer registry with underlying pattern is determined by lattice misfit, dewetting is favored for strongly interacting particles, and sparse nanoparticle arrays may be patterned by using a curved or non-close-packed base layer.

1.5.2 *Temperature*

Tuning solution thermal energy offers another route to control phase behavior of colloidal nanoparticles. For example, even in a good solvent for ligands, formation of small nanoparticle clusters is common, particularly for strongly interacting (e.g. metal) core materials with high-curvature surfaces (i.e. small-diameter nanoparticles).[95] Because osmotic pressure from corona overlap increases with temperature, heating nanoparticle solutions encourages dispersal, and cooling induces flocculation.

Along these lines, controlled chilling of nanoparticle solution is, to our knowledge, an unexplored technique for self-assembly, and might be expected to produce superlattices similar to the non-solvent destabilization approach. On the other hand, tuning the temperature of evaporating nanoparticle solution has been shown to have a strong effect on superlattice phase behavior.[88] For example, spherical nanoparticles evaporated from low temperature solution assemble into non-compact superlattices, while at elevated temperatures close-packed phases are observed. Furthermore, the clustering of strongly interacting nanoparticles in solution at room temperature favors the formation of cluster-incorporating (e.g. $NaZn_{13}$ or $CaCu_5$) binary nanoparticle superlattices (BNSLs), while elevated temperatures promote binary phases featuring metal nanoparticles occupying interstitial sites more evenly distributed (Figure 1.6).

Thermal annealing can induce superlattice phase transitions even long after solvent drying. For example, Korgel and co-workers demonstrated that heating of monodisperse close-packed Au nanoparticles up to above 200°C can result in Ostwald ripening, yielding bidisperse nanoparticle ensembles that self-assemble in the solid state into pseudo-Frank–Kasper $CaCu_5$ and $NaZn_{13}$ BNSLs.[96] In a similar follow-up work, Pileni and co-workers observed thermal-ripening-induced solid state assembly at temperatures in the range 25–130°C into, among other phases, the Frank–Kasper $MgZn_2$ structure.[97]

Such thermal coalescence of nanoparticle cores and concomitant structural changes are irreversible. On the other hand, fully reversible thermal toggling of superlattice structures was recently demonstrated by the Korgel group. In this intriguing experiment, a disordered solid of slightly polydisperse, octadecanethiol-capped Au nanoparticles was coaxed into adopting a body centered cubic (bcc) lattice at moderately elevated temperatures (60°C), while ordering was diminished again upon cooling back to room temperature.[98] The authors proposed that the Au core polydispersity can be softened by the molten nanoparticle corona, facilitating ordering of the nanoparticles above the melting temperature of octadecyl chains. Freezing of chains upon superlattice cooling, however, induces formation of hydrocarbon bundles and prevents the necessary uniform distribution of segments throughout the superlattice matrix. Interestingly, the opposite trend is observed for polydisperse Ag nanoparticles capped with end-functionalized liquid crystal molecules: bcc superlattices undergo a reversible order–disorder transition order upon heating to 120°C.[99]

1.5.3 *External fields*

Another means of directing self-assembly of colloidal nanoparticles involves the use of external fields.[100] In this case, the particles experience a rotational force

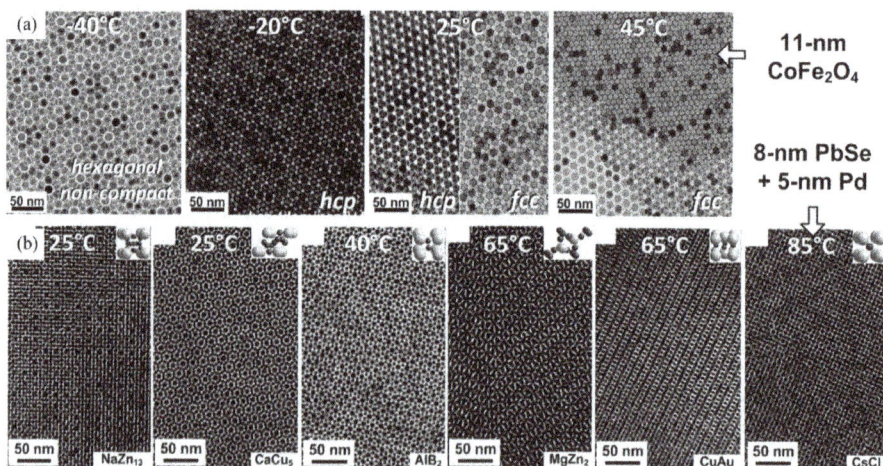

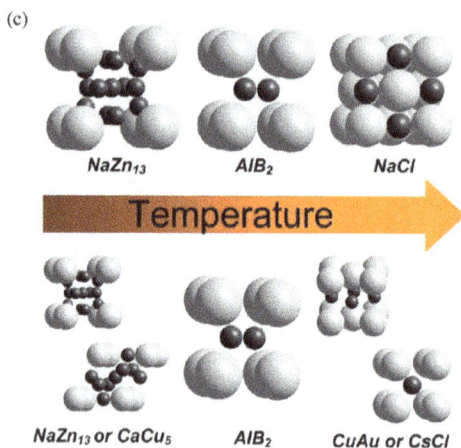

Figure 1.6   Temperature-dependent phase behavior of spherical nanocrystals. (a) Evaporating a solution of 11 nm $CoFe_2O_4$ nanocrystals at –40°C yields non-close-packed superlattices, while increasing the solution temperature during evaporation yields hcp or fcc arrangements. (b) Evaporating a binary solution of 7.7 nm PbSe and 4.9 nm Pd (γ_{eff} = 0.74) yields cluster phases ($NaZn_{13}$, $CaCu_5$) at low temperature and interstitial phases (CuAu, CsCl) at higher temperature. (c) Schematic illustration of the effect of temperature tuning on the binary phase behavior of a semiconductor-semiconductor system (top, TEMs not shown) and a semiconductor-metal system (bottom, corresponding TEMs shown in (b)). Adapted from Ref. 88. Copyright 2010 American Chemical Society.

(torque) favoring parallel alignment of the nanoparticle permanent electric or magnetic dipole moment with the external field. Evaporating a solution of nanoparticles in the presence of such a field condenses prealigned particles into superlattices with orientational registry of dipole moments. For example, drying a toluene solution of CdS nanorods spread between electrodes produces single-layer and multilayer smectic B liquid crystals with the rod long axis along the direction of the applied electric field.[101]

The application of an external field also aligns flat surfaces of the rods, facilitating close-packing. Micrometer-sized polyhedral metal–organic framework crystals align in an electric field into only 1D locked chains that otherwise remain uncorrelated,[102] an observation distinct from their behavior in the absence of the external field.[103] Furthermore, particles have been guided into specific locations for assembly, for example *via* the use of positive Alternating current (AC) dielectrophoresis into narrow gaps,[104] a phenomenon also observed in simulation.[105] With a spatially varying magnetic field Fe_3O_4 nanoparticles were aligned into parallel line patterns on the surface of a disk medium.[106]

When the nanoparticle dipole moment and close-packing direction are not aligned, complex superstructures can develop. For example, evaporation-based assembly of cube-shaped 13-nm Fe_3O_4 nanoparticles in the presence of an external magnetic field yields 1D superlattice helices.[107] Such braided, chiral structures appear to be a natural compromise between the system's entropic preference for face-to-face packing of cubes and the energetic drive to align the nanoparticle magnetic easy axis (cube diagonals) with the long dimension of the 1D superstructure (Figure 1.7). Similar braided structures have also been found to assemble

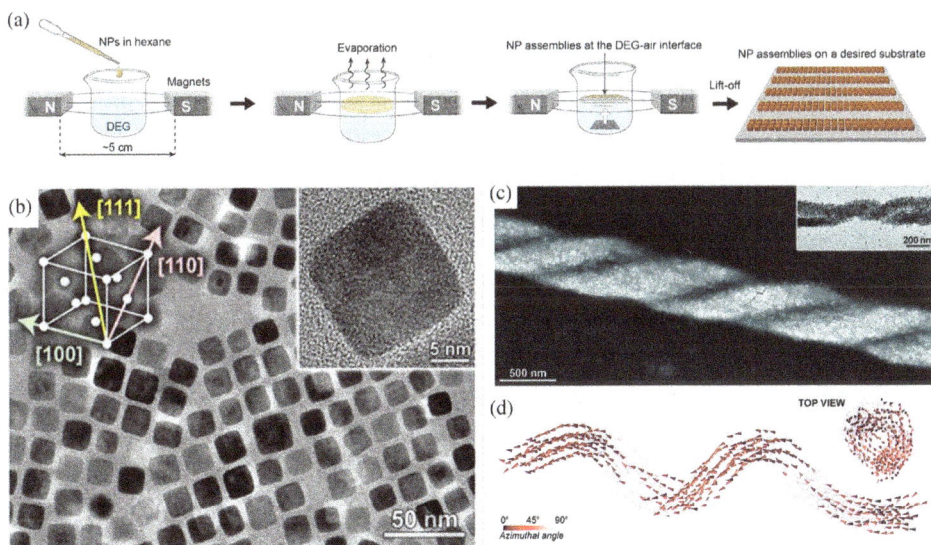

Figure 1.7 Self-assembly of magnetic nanocubes in external magnetic field. (a) Schematic of experimental setup, whereby a hexane solution of nanocrystals is evaporated between two magnets. (b) TEM image of 13-nm Fe_3O_4 rounded nanocubes used in assembly. The (111)-, (110)-, and (100)-crystallographic axes are easy, intermediate, and hard axes of magnetization, respectively. (c) SEM image of double helix, with TEM image inset. (d) Snapshots from Monte Carlo simulations of a 1D belt folding into a helix. Adapted with permission from Ref. 107. Copyright 2014 American Association for the Advancement of Science.

from aqueous dispersions of 4-nm CdTe truncated tetrahedral passivated with thio-glycolic acid. In such systems, exposure to ambient light leads to photooxidation and attachment of particle surfaces, ultimately producing twisted ribbons with equal distribution of left- and right-handed chirality.[108] On the other hand, when a racemic dispersion of such particles is illuminated by circularly polarized light, enantioselective photoactivation takes place, resulting in the formation of chiral twisted ribbons with about 30% enantiometric excess.[109]

When the induced magnetic dipolar interaction and the interaction between the magnetic dipoles of cubes and the external field are small, on the order of the thermal energy and van der Waals forces, then other structures such as chains, ribbons, sheets, and large cuboids have also been observed.[110–112] Buckling of chains and cross-linking occurs only at sufficiently strong fields.[113,114] Along these lines, fluctuating electromagnetic fields can induce a van der Waals torque,[115] which leads to oriented attachment of cubic or spherical $BaTiO_3$ nanoparticles.[116] Interestingly, light-controlled self-assembly is possible using non-photo responsive nanoparticles by using a photo switchable medium that responds to light in such a way that it modulates the interparticle interaction.[5,117]

The ability of an external field to direct the alignment of nanoparticles is especially beneficial for the assembly of anisotropic nanoparticles, which requires orientational ordering aside from positional ordering of individual building blocks. Under the influence of an external field, 1D nanoparticles align with their longitudinal axis oriented along the direction of the field lines.

For colloidal 1D nanoparticles that carry a net charge and/or a permanent electric dipole moment, an electric field is the perfect external stimulus for directing their assembly.[101,118–120] In the presence of an applied electric field (E), alignment is achieved when the total electric dipole moment ($d = d_1 + d_0$, where d_1 and d_0 are the induced and permanent electric dipole moment, respectively) gives an alignment energy ($U_{align} = E \times d$) that is strong enough to overcome the room temperature thermal excitation energy ($k_B T = 26$ meV, where k_B is the Boltzmann constant) that would otherwise randomize the dipole orientation.[120] Thus, there is a minimum value of electric field strength (E_{min}) that is necessary for significant alignment to occur. Figure schematically show examples of experimental setups that have been used in aligning colloidal nanorods of wurtzite-phase cadmium chalcogenides under direct current (DC) electric fields. In general, a non-conducting flat substrate (e.g. Si_3N_4) is positioned between parallel electrodes that are arranged either in a coplanar manner or in a top-down manner.[120] The colloidal nanorod dispersion is deposited onto the substrate and the solvent is slowly evaporated over a time period (usually several hours) during which a direct voltage of sufficient magnitude is applied.

Binary cadmium chalcogenides possess a non-centrosymmetric wurtzite lattice that gives rise to a permanent electric dipole moment that increases with the nanorod volume.[119] When placed under the influence of an external electric field, the nanorods experience a torque that rotates them to align their long axis with the field direction (i.e. perpendicular to the faces of the electrodes). The rods can align over large areas by following the field streamlines throughout the electrode gap.[118] The placement of the electrodes determines the field direction and thus dictates the final orientation of the nanorods with respect to the substrate that is between the electrodes. When the electrodes are arranged in a coplanar manner, the rods align in the plane of the substrate as shown in Figure. By contrast, a top-down arrangement of the electrodes results in nanorods that are oriented perpendicular to the substrate as seen in Figure. Meanwhile, controlled evaporation of solvent assists in the close-packing of the oriented nanorods and the degree of positional order is dependent on the evaporation rate. A higher degree of positional order is observed for a slower rate of solvent evaporation.[101]

AC electric fields can also be employed to direct the formation of 1D nanoparticle assemblies using similar experimental setups as those described above but with alternating voltage applied to the electrodes. An advantage of using an AC electric field is that it avoids the interference of electro-osmotic and electrochemical effects that are present when DC is used.[57] Mayer and co-workers have demonstrated the assembly of Au nanowires dispersed in a dielectric medium upon application of an AC electric field using coplanar interdigitated electrodes.[121] The metallic nanowires are easily polarized in the electric field due to charge separation at the nanowire surface. The nanowire alignment process can be influenced by the magnitude as well as the frequency of the alternating voltage. A shorter alignment time is observed when the magnitude and frequency of the alternating voltage are increased.

For anisotropic nanostructures that are magnetically active (i.e. with permanent or field-induced magnetic dipole moment), an external magnetic field can be utilized to facilitate self-assembly by forcing these building blocks to align with their long axis parallel to the magnetic field lines.[122] The position of the magnet relative to the substrate dictates the orientation of the particles whereas the strength of the magnetic dipolar interactions between the particles influences the positional order.[123] The magnetic field strength can be adjusted by altering the distance between the magnet and the dispersion. Song and co-workers have demonstrated the use of an applied magnetic field (produced by a RuFeB magnet) to control the orientation of ellipsoidal core–shell γ-Fe$_2$O$_3$-SiO$_2$ particles during solvent evaporation.[124] A close-packed 3D array of ellipsoidal particles with both orientational and positional order was realized. For nanorods and nanowires, magnetic facilitations have been shown to produce side-by-side (e.g. FePt nanorod rail-tracks)[125] and end-to-end (e.g. Ni nanowire chains)[126] self-assemblies.

1.6 Superlattice of Nanoparticle with Different Shape

In the following section, we highlight examples of superlattices formed by spherical and anisotropic nanoparticles and nanoparticles mixtures capped with aliphatic surface ligands. We connect the observed superlattices to the physical concepts underlying the self-assembly of hard shapes and soft shapes. An overview of self-assembled structures obtained for various nanoparticle shapes is provided in. Due to high chemical potential of high-curvature surfaces, anisotropic nanoparticles typically have rounded or truncated corners. Descriptions of an inorganic core shape as cube, octahedron, and tetrahedron should therefore be understood as slightly truncated/rounded versions of the shape.

1.6.1 *Spherical Nanoparticle*

Self-assembly of nearly spherical (e.g. cuboctahedral, rhombicuboctahedral, or multiply twinned icosahedral)[127] nanoparticles by solvent evaporation often results in the formation of fcc or hcp superlattice thin films. Because superlattice domains are typically oriented with the close-packed plane making contact with the support, fcc and hcp phases are often observed to assemble with the 3-fold axes parallel to the normal of the thin film. These phases are built up from hexagonally ordered monolayers with ABCA- or ABAB-type overlay pattern. In line with hard sphere phase expectations, fcc and hcp are most common for particles with surface ligands that make a minimal contribution to total particle volume (small softness value L/R). The small free energy difference between fcc over hcp predicted for hard spheres appears to be negligible in nanoparticle self-assembly experiments as both phases are commonly produced (Figure 1.8).[128]

However, fcc and hcp are not always observed with equal probability. Experiments with 5-nm-diameter alkanethiol-capped Ag nanoparticles indicate that evaporation-based assembly from higher-boiling solvents (e.g. octane) favors hcp structure, while lower-boiling solvents (e.g. hexane) typically produce fcc superlattices.[129] Along these lines, it has been hypothesized that solvent flow through the interstitial sites of close-packed spheres may play a role in favoring fcc packing by directing the third layer of spheres toward the threefold sites open to solvent flux.[130,131] Furthermore, dipole–dipole interactions between spherical nanoparticles favor anti-ferroelectric ordering of dipole moments and promote formation of phases that maximize dipolar energy. As a result, the alignment of

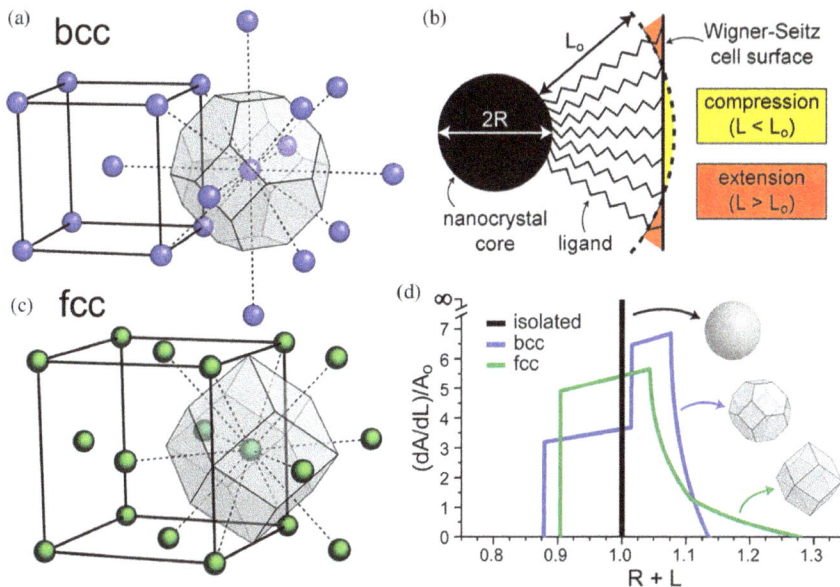

Figure 1.8 Wigner–Seitz evaluation of sphere packings. The bcc structure requires less distortion of the ligand corona than fcc arrangement. (a) bcc unit cell and the corresponding Wigner–Seitz polyhedron. (b) bcc unit cell and the corresponding Wigner–Seitz polyhedron. (c) Illustration of the compression and extension of hydrocarbon chains required to occupy the entire Wigner–Seitz cell. (d) Asphericity of both Wigner–Seitz cells. The spread in center-to-surface distances of the bcc Wigner–Seitz polyhedron (blue trace) is narrower than that of the fcc Wigner–Seitz polyhedron (green trace), and thus requires less distortion of capping ligands. Adapted from Ref 128. Copyright 2015 American Chemical Society.

such particles along columns with head-to-tail dipole–dipole orientation can stabilize hcp or even simple hexagonal phases over fcc.[132]

Subtle differences in the shape of quasi-spherical metal nanoparticles influences the presence or absence of orientational registry in self-assembled superlattices. For example, typical syntheses of Au nanoparticles often result in a mixture of single-crystalline (cuboctahedral) and polycrystalline (multiply twinned icosahedral) nanoparticles. Slow destabilization of such colloids results in phase separation of the two components: the more spherical polycrystalline icosahedral nanoparticles assemble into rotationally degenerate fcc superlattices, while the larger facets of crystalline cuboctahedral nanoparticles direct formation of fcc superlattices with orientational registry between Au nanoparticle cores.[133] But core–core registry need not result in identical nanoparticle orientations. Detailed structural analysis of fcc superlattices of PbS nanoparticles indicates that particles occupying corner sites of the unit cell can be oriented differently than those occupying the face-centered sites.[134]

Nanoparticles with significant soft character often assemble into bcc superlattices following the prediction of the soft particle model. The well-established phase behavior of alkanethiol-capped Au nanoparticles, for example, shows a critical softness value that serves to differentiate fcc from bcc forming Au nanoparticles at $L/R \approx 0.7$. Along these lines, installation of octadecanethiol (C18) chains on 2-nm-diameter spherical Au nanoparticles ($L/R \approx 1$) yields bcc superlattices, while the same ligands on 5-nm-diameter spherical Au nanoparticles ($L/R \approx 0.5$) yields hcp structures. Such observations are consistent with the proposition that the geometry of the space available to a particle in the fcc/hcp arrangement deforms surface hydrocarbon chains more significantly than that of the bcc lattice, destabilizing fcc/hcp arrangements of nanoparticles with significant contribution from the soft ligand shell.

1.6.2 *Nanorod*

Rod-shaped nanoparticles synthesized by colloidal chemistry approaches enable exploration of the self-assembly behavior of anisotropic particles. Besides the isotropic liquid (random position and orientation) and the crystalline solid (fixed position and orientation), there are two families of mesophases, or states of matter with characteristics of both solid and liquid phases. Rod-shaped particles can adopt nematic (random position, fixed orientation) and smectic (fixed position in a plane only, fixed orientation) packings. In close analogy, platelet-shaped disks adopt nematic and discotic columnar (fixed position along one axis only, fixed orientation) mesophases.

Colloidal chemistry techniques have produced uniform collections of nanorods and nanowires which assemble into a number of ordered structures.[135,136] For instance, metal (e.g. Au and Ag) nanorods grow from seeds in the presence of cationic surfactant (e.g. cetyltrimethylammonium bromide, CTAB) micelles.[137,138] In such systems, optimizing surfactant mixtures have proven crucial to suppressing undesired shape byproducts. Exceptional size and shape uniformity of Au nanorods have recently been achieved using CTAB and sodium salicylate,[28] CTAB and sodium oleate,[139] and alkyltrimethylammonium chloride and sodium oleate.[140] Furthermore, semiconductor (e.g. CdSe, CdSe–CdS) nanorods can be prepared *via* high-temperature precursor decomposition in non-polar solution.[1,30] In this case, rapid monomer incorporation along weakly passivated nanoparticle surfaces induces unidirectional growth along a particular crystallographic axis. More recently, synthetic routes to 2D semiconductor (e.g. CdS, CdSe, CdTe)[141] and various oxide and fluoride (e.g. UO_2, LaF_3, etc.) nanoplates[142] have been developed.

Evaporating a solution of colloidal nanorods yields superlattices in which the rods are highly aligned but exhibit some disorder in their position of rod centers.

Layers of similarly oriented rods (lamellae) often contain hexagonally close-packed rods (smectic B phase). Optical micrographs of nanorod liquid crystals show schlieren textures and disclinations common to conventional organic molecular liquid crystals.[143] CTAB-capped Au nanorods sometimes assemble tip-to-tip in a comparatively open superlattice structure not anticipated by simulations of hard rods. This behavior can be explained by recent *in situ* TEM experiments,[144] which provided evidence for anisotropic repulsion interaction resulting from reduced charge density at the tips of CTAB stabilized Au nanorods in water. Self-assembly of uncharged nanorods (e.g. CdSe–CdS dot-in-rods capped with octadecylphosphonic acid, ODPA, ligands) by solvent evaporation produces smectic B superlattice films with either horizontal or vertical alignment depending on the choice of subphase.[145] The interplay between hexagonal close-packing within rod lamellae and the hexagonal cross section of wurtzite CdS rods facilitates orientational registry of the rods about their symmetry axes, maximizing cohesive interactions between flat rod faces after solvent evaporation.

Destabilization of the colloidal solution by non-solvent addition,[71] depletant addition,[51] or micelle decomposition (Figure 1.9),[74,146] offers alternative routes to

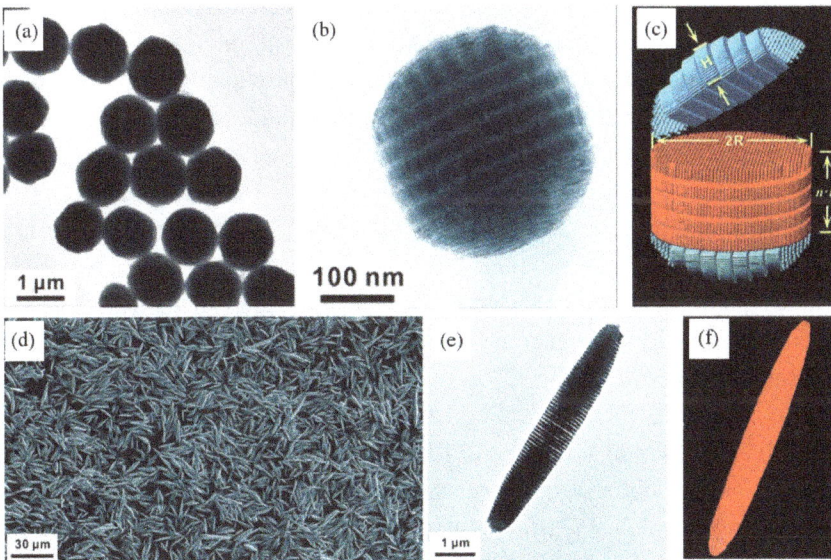

Figure 1.9   Superlattices formed by destabilization of nanorod colloids. (a) TEM overview of nearly spherical micrometer-sized superparticles comprised of CdSe–CdS dot-in-rods. (b) TEM zoom and (c) model of the superparticle microstructure revealing that rods are packed into a central cylindrical domain capped on either end by a multilayer dome. (d) TEM overview of single-domain elongated needles formed by preferential rod–rod attachment *via* highly solvophobic tips. (e) TEM zoom and (f) model of superparticle needles. Adapted with permission from Ref. 74. Copyright 2012 American Association for the Advancement of Science.

novel morphologies of rod assemblies. In one notable example, Cao and co-workers demonstrated that CdSe–CdS nanorods capped with ODPA surface ligands and water-solubilized by overcoating with DTAB surfactants controllably aggregate in ethylene glycol into nearly spherical or needle-shaped micrometer-sized colloidal superparticles. Detailed characterization revealed that the superparticles with >80,000 rods are comprised of a central cylindrical domain of close-packed rods capped by two domes. The amount of DTAB present in the aqueous nanorod solution controls the size of superparticles. An embryo growth mechanism was proposed, whereby small rod domains below the critical radius of stable nuclei coalesce into the capped cylinder shape. On the other hand, destabilization of the colloid by gradual diffusion of polar non-solvent produces platelet shaped nanorod assemblies. In addition, flat sheets of hexagonally packed nanorods have been prepared by oleic acid addition, triggering nanorod assembly *via* depletion attraction.

1.6.3 *Nanoplates*

Thermal decomposition of precursors in coordinating solvents has been applied to prepare 2D nanoparticles (nanoplates) from cadmium chalcogenides and rare-earth halides.[141] For colloidal nanoplates lacking uniform lateral dimensions, both solvent evaporation and destabilization approaches produce face-to-face stacks of ribbon-like 1D superstructures.[72,147,148] On the other hand, nanoplates with uniform size and shape, including circles, hexagons, rhombi, ellipses, and tripods, assemble into 2D and 3D superlattices. Slow evaporation of dilute solutions of nanoplates at the liquid–air interface induces self-assembly into area-tiling superlattices with the plate c-axis perpendicular to the support. Because the plate footprint determines the densest tiling arrangement, varying platelet shape leads to a variety of 2D patterns including hexagonal, square, and rhombic tilings as confirmed by simulation and experiment.[149]

While entropy is a main driving force in larger, micrometer-sized colloidal plates such as Brownian squares and triangles,[150,151] self-assembly of nanometer-sized plates can be understood from maximization of tiling density only to first approximation. For the case of irregular hexagonal DyF3 nanoplates, comprised of four (101) edge ("A") facets at the tips and two (002) side ("B") facets in the middle, an alternating arrangement resembling the herringbone packing is formed. Geometric analysis reveals that while the alternating arrangement perfectly tiles a surface at intermediate plate aspect ratio (edge lengths, B/A), the parallel arrangement is area-tiling regardless of aspect ratio. Because the alternating arrangement never covers a surface more efficiently than the parallel arrangement, this system requires an explanation beyond the entropy-maximizing perspective of efficient polygon packing. In fact, density functional theory calculations presented in the same work revealed that oleic acid surface ligands bind strongly to the (101) tip edge (A) facets and weakly to the (002) middle edge (B) facets. Assuming edge–edge attractions arising from ligand vdW after solvent evaporation increases with the number of interacting alkyl chains, in-plane cohesion energy is greater for the alternating plate arrangement that avoids weakly interacting B–B contacts. Indeed, Monte Carlo simulations incorporating such an edge-specific interaction asymmetry predict a stability range for the alternating arrangement.

In addition to tiling a flat surface, colloidal nanoplates also self-assemble into 3D superlattices. Evaporating a hexane solution of rhombic GdF3 nanoplates over glycol liquid support yields either columnar or lamellar structures. Both columnar and lamellar plate phases feature face-to-face contacts between plates, an observation that may be rationalized based on Onsager-type entropic arguments as well as maximization of cohesive van der Waals interactions between the plates after solvent drying. 1D lamellar superlattices of triangular Au nanoprisms featuring anomalously large interplate separations have been rationalized by treating the CTAB surfactant contribution to both attractive depletion forces and repulsive electrostatic forces.[77]

1.6.4 *Polyhedral Particle*

The body of knowledge collected by Monte Carlo simulations of phase behavior of hard polyhedra serves as a logical starting point for considering the self-assembly of highly faceted colloidal nanoparticles with regular polyhedral shape. Maximization of translational entropy in solution and interparticle cohesion after solvent evaporation both favor formation of dense polyhedron packings with a preference for face-to-face contacts. Rotational entropy, on the other hand, favors structures that assemble polyhedra in many orientations but leave space for particle reorientations. Indeed, Yang and co-workers demonstrated that 500 nm Ag polyhedra with adsorbed poly(vinylpyrrolidone) (PVP) ligands sediment from methylformamide (good solvent for PVP) under gravity to form densest packings known for cubes, cuboctahedra, and octahedral (Figure 1.10).[152] Similarly, 10 nm cubes of semiconductor,[153] metal[154,155] and metal oxide[107,156] materials form close-packed arrangements with face-to-face contacts and cubic superlattice symmetry. Metal oxide octahedra (edge length about 35 nm) assemble into the densest packing available to the shape,[157] and metal oxide tetrahedra with similar edge lengths also readily establish face-to-face contacts.[158] Even highly complex phases like the network phase isostructural with high-pressure

Figure 1.10   Self-assembly of densest polyhedron packings by gravitational sedimentation of polymer-stabilized, 500 nm Ag nanocrystals. (a)–(e) SEM images (left) and modeled structures (right) of dense polyhedron (cube, truncated cube, cuboctahedron, truncated octahedron, octahedron) packings. Adapted with permission from Ref. 152. Copyright 2011 Nature Publishing Group.

lithium found for weakly rounded octahedral can be explained within the hard particle model.[159]

However, the entropy maximization principle is not sufficient to predict the self-assembly behavior of smaller particles. Octahedra and tetrahedra with comparatively short edge lengths on the order of 10 nm show phase behavior distinct from the expectations set by hard-particle studies and distinct from observed superlattices of analogous particles with size on the order of 100 nm. CdSe tetrahedra and Pt_3Ni octahedra in this size regime form superlattices with tip-to-tip contacts.[160,161] Such tip-to-tip contacts preserve avoidable void space in the lattice, which is disfavored for hard object packings.

The superlattices of smaller polyhedral nanoparticles can be rationalized by considering the influence of surface curvature on repulsive ligand–ligand interactions during evaporation-based self-assembly. As nanoparticles are condensed into a small volume of solvent, overlapping ligand coronas experience unfavorable accumulation of hydrocarbon segments. The resulting repulsion energy depends on the geometry of the nanoparticle contact. Overlap of face-bound ligands results in stronger spatial concentration of chain segments than overlap of ligands tethered to curved nanoparticle surfaces. Accordingly, the osmotic penalty associated with unfavorable exclusion of good solvent gives rise to a significant difference in per-chain repulsion energy between face- and vertex-bound ligands favoring contact between nanoparticle tips in the crowded assembly solution. Besides this energetic argument, the effective rounding of vertices and edges decreases the facet area that can be in contact, weakening directional entropic forces. In addition, the sacrifice of two of the three rotational degrees of freedom upon establishment of face–face contacts further penalizes arrangements incorporating such geometry. Energetic and entropic factors can thus favor superlattice structures not anticipated by hard polyhedron packing arguments.

Just as in the case of rods, application of the destabilization approach to assembling polyhedral nanoparticles leads to an intriguing set of superlattice shapes from nanoparticles with cubic, rectangular, rhombic dodecahedral, octahedral, and truncated octahedral core geometries.[156,162,163] For example, aqueous precipitation of 40 nm polyhedral Au nanoparticles passivated with alkylammonium halide surfactant bilayers results in micrometer-sized superlattices with morphology that varies systematically with nanoparticle shape.[164]

1.6.5 *Binary Nanostructures*

In addition to superlattices formed by self-assembly of size- and shape-uniform nanoparticles, co-crystallization of two nanoparticle species is also possible.[165] Micrometer-sized spherical colloids have been shown to assemble, depending on size ratio, into a few phases of 3D and 2D binary superlattices that may be rationalized based on sphere packing principles.[166,167] However, as in the case of octahedra, simple hard-shape phase behavior observed for micrometer-sized particles gives way to more complex superlattice structures at the sub-20-nm length scale. Along these lines, Murray and co-workers observed more than 15 unique BNSLs by evaporating a solution containing two sizes of quasi-spherical nanoparticles about a decade ago.[168] Since then, the number of unique binary structures has continued to grow. Recent additions are a quasi-crystalline BNSL,[169,170] one isostructural with $C_{60}K_6$ (also called bcc-AB_6),[171] one with A_6B_{19} stoichiometry,[172] and the Li_3Bi structure.[173]

Binary sphere mixtures often pack more densely than a single component alone, for example, by filling the voids in a close-packed sphere lattice with smaller spheres. When evaluating ways to densely pack sphere mixtures, there are two degrees of freedom that influence the maximally achievable packing density: the radius ratio (or size ratio), $\gamma = R_B/R_A$, and the stoichiometry, $\gamma = n_B/(n_A + n_B)$, of the large (A) and small (B) spheres. Recent studies uncovered more than 15 unique binary sphere packings that exceed the densest single-component (fcc) arrangement.[174] Such analyses provide a natural starting point for anticipating the structures formed by spherical nanocrystals that seek to maximize packing density

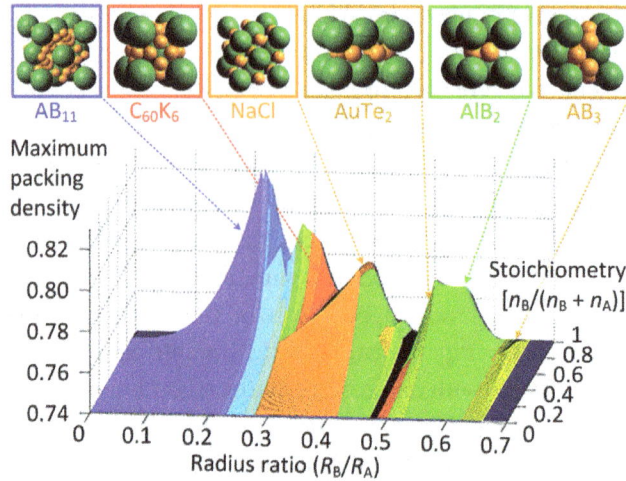

Figure 1.11 Summary of densest known binary sphere packings. Maximum packing density (z-axis) surface plot shown as a function of radius ratio (x-axis) and stoichiometry (y-axis). Unit cells or characteristic structural motifs for selected structures are shown above. The proposed maximum density is claimed by the AB11 structure, which fills space with about 82% efficiency at radius ratio close to 0.22. The radius ratio for which binary packings exceed single-component close-packing is $\gamma < 0.66$. Adapted with permission from Ref. 174. Copyright 2012 American Physical Society.

at high particle volume fraction. However, in the limit of similar sphere radii ($\gamma > 0.66$), phase separation into separate fcc (or hcp) lattices of large and small spheres provides the densest packing, while in the limit of very disparate sizes ($\gamma < 0.2$) depletion effects strongly disfavor the achievement of dense packings in experiment (Figure 1.11).

More recent variations on this theme include ternary nanoparticle superlattices,[175] 2D BNSLs,[176] BNSLs assembled from polystyrene-capped nanoparticles,[177] and BNSLs incorporating polyoxometallate clusters.[178] Such diversity is often observed even within a single sample, which may present a few different binary structures together with phase-separated domains (e.g. ABm + ABn + fccA + fccB). In Figure, we show 16 unique BNSL structures assembled from quasi-spherical hydrocarbon- capped nanoparticles, and in Table we tabulate, to the best of our knowledge, all binary structures observed to date. Although the stability of some BNSL phases (e.g. NaCl, AlB$_2$, C$_{60}$K$_6$) can be rationalized as dense packings of a mixture of two sizes of spheres, most structures observed to date have lower sphere packing density as compared to the phase separated fccA + fccB alternative. The treatment of nanoparticle self-assembly as a packing problem is exact only in the high-pressure limit, however. At finite pressure (i.e. intermediate particle volume fractions), the configurational contribution to total system entropy can stabilize lower-density phases with large and complex unit cells such as $NaZn_{13}$ and $MgZn_2$. Thus, if the observed (dry) BNSL structures are a reflection of the stable phase in the colloidal (wet) crystal, only those with density above approximately 0.65 might be rationalized based on hard-sphere interactions. Particularly for sub-10-nm nanoparticles, the application of packing arguments is further complicated by the presence of the hydrocarbon ligand shell. The ligand shell, which can make a significant contribution to the total particle size, can be accounted for by taking the effective size ratio to be the quotient of the effective diameters of large and small nanoparticles measured from center-to-center separations in phase-separated, close-packed films. The experimentally observed effective size ratio can be compared to the size ratio calculated as $\gamma_{eff} = (D_B + 2L_B)/(D_A + 2L_A)$. D_A (L_A) and D_B (L_B) are the core diameters (effective ligand thicknesses) of large and small nanoparticles, respectively. However, even with this adjustment, experimental data conflicts with the stability range predicted from the hard sphere model. For example, $NaZn_{13}$ and $MgZn_2$ structures are observed across radius ratio ranges ($0.46 < \gamma_{eff} < 0.74$ and $0.65 < \gamma_{eff} < 0.81$, respectively). These ranges lie outside the boundaries for which these hard sphere phases are predicted to be stable ($0.54 < \gamma < 0.61$ and $0.76 < \gamma < 0.84$, respectively).

It is tempting to explain the discrepancy between experimental BNSL phases and hard sphere predictions by considering the nanoparticle ligand corona. For example, particle tracking measurements of single-component and binary nanoparticle arrays reveal changes in the effective capping layer thickness for nanoparticles in various coordination environments.[173] From this data, it was concluded that a difference in hydrocarbon segment density across asymmetric brush contacts leads to corona deformation. The deformation allows the softer component to fill space more efficiently than a rigid sphere placed in the same lattice site, increasing the density of the binary structure beyond that possible within the hard sphere approximation. Furthermore, unanticipated BNSL structures may also incorporate area-minimizing character that minimizes elastic deformation of surface bound hydrocarbon chains by providing particles with more spherical Voronoi cells than those offered by the set of densest sphere packing. Indeed, the observation of tetrahedrally close-packed (tcp) structural motifs in several common BNSLs including Frank–Kasper $MgZn_2$ and pseudo-Frank–Kasper $CaCu_5$ and $NaZn_{13}$ phases is an indication that there exists some area-minimizing component to BNSL assembly.

Thermodynamic calculations confirm that short-range, soft repulsive potentials enrich the phase diagram beyond that of hard spheres by stabilizing many common BNSL phases including AlB_2, $MgCu_2$, $CaCu_5$, $C_{60}K_6$, and $NaZn_{13}$.[179] Such soft repulsive potentials, explored in detail as effective interactions for star polymers and other soft matter systems, mimic ligand–ligand steric interactions well.[180]

Energetic interactions between particles may also contribute to the formation of BNSL structures unanticipated for systems dominated by hard-sphere phase behavior. For example, self-assembly of PbSe and Pd nanoparticles from solutions between −20 and 85°C was observed to produce seven different BNSLs across the temperature series.[88] In this study, the BNSLs observed at low temperatures frequently incorporated clusters of metal nanoparticles (e.g. $CaCu_5$, $NaZn_{13}$), suggesting that co-crystallization under such conditions proceeds by integration of preassembled clusters of strongly interacting metal nanoparticles. In addition, electrostatic charging has also been implicated as a potential source for rich binary phase behavior.[181] Indeed, complex binary structures reminiscent of BNSLs were observed in ionic colloidal crystals of oppositely charged particles,[182] and similarly rich binary phase diagrams have been calculated with addition of electrostatic interactions between nanoparticles.[181] Even so, the poorly screening non-polar solvents in which hydrocarbon-capped nanoparticles are prepared and assembled (e.g. octadecene, octane, toluene, etc.) should be an unlikely environment to observe charge separation.

The preceding analysis reveals that, 10 years after its revelation, the surprising structural diversity of BNSLs has yet to be fully explained. Nevertheless, progress has been made toward uncovering potential sources of counterintuitive phase behavior including ligand corona deformability, elastic preference for area-minimizing lattices, and kinetic integration of preformed clusters. Combining any or all of the above effects with established principles of hard-sphere crystallization may be necessary to achieve de novo BNSL structure prediction.

1.7 Applications of Assembled Structure

Nanoparticle self-assembly has presented researchers with several intriguing puzzles. Explaining the diversity of the observed structures made from nanoparticles requires a rationale that invokes a combination of hard and soft matter ordering principles as well as kinetic and environmental effects. Beyond the fundamental questions, however, lies a practical motivation: the structural hierarchy and compositional tunability of nanoparticle superlattices provides these materials with novel electronic, optical, mechanical, and catalytic functionality for nanotechnological applications. Realizing this potential remains the ultimate goal of nanoscience.[7,183,184] Just as the factors governing crystal structures of bulk solids (e.g. atomic radius, chemical valence) are inextricably linked to properties of practical importance (e.g. electron affinity, ionization energy, polarizability), controlling the position, orientation, and chemical composition of nanoparticle components thus promises a versatile design platform for custom high performance materials.

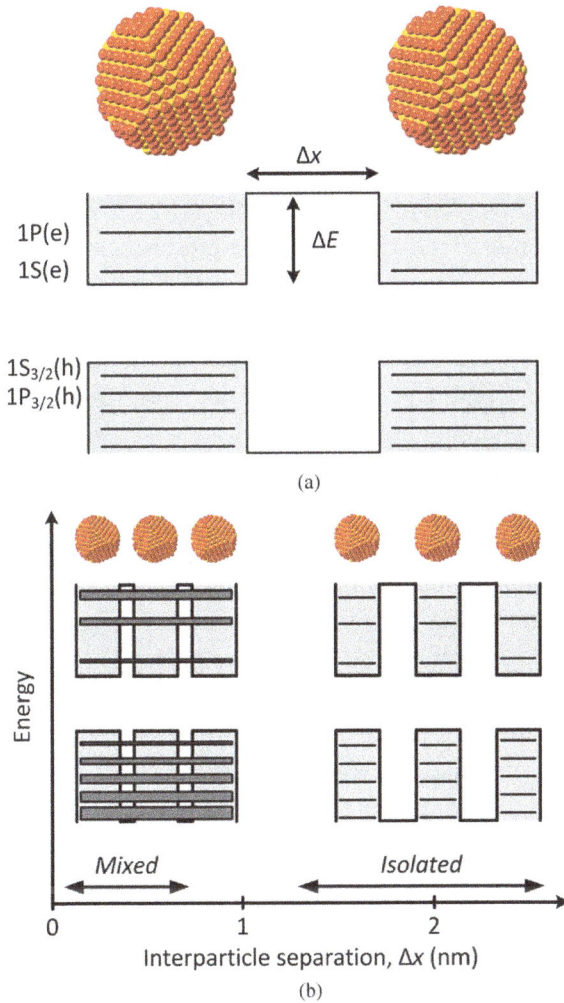

Figure 1.12 Control over the development of collective electronic states. (a) Sketch of two CdSe nanocrystals (top) and their quantized electronic states (bottom) separated by an interparticle matrix of barrier height ΔE and surface-to-surface distance Δx. (b) Schematic illustration of the spatial extension of electronic wave functions upon reduction in Δx. Adapted with permission from Ref. 186. Copyright 2015 Nature Publishing Group.

1.7.1 *Optoelectronic*

Semiconductor nanoparticle superlattices, for example from metal chalcogenide nanoparticles (quantum dots),[185] are candidates for materials with optoelectronic applications such as light-emitting devices (LEDs), photodetectors, and solar cells.[186] Mixing of electronic states between nanoparticles, analogous to the development of molecular orbitals by overlap of atomic wave functions is crucial for this purpose. Unfortunately, superlattices of nanoparticles with traditional hydrocarbon stabilizers cannot be directly used. The ligand bilayers separate the nanoparticles too far and act to electronically isolate neighboring particles from one another, creating a high tunnel barrier for charge transport among neighboring nanoparticles. As a result, the electronic and optical properties of the superlattices are often similar to those of the isolated building blocks. Weak coupling gives rise to Anderson localization,[187] contributing to the presence of mid-gap electronic trap states. Stronger electronic coupling between nanoparticles is expected to improve charge carrier mobility by suppressing domain localization and may reveal experimental signatures for new collective phenomena in nanoparticle assemblies obscured so far by averaging over a range of coupling strengths.

A reduction in tunneling barrier height has been achieved by exchanging the hydrocarbon ligands used for synthesis with compact, nucleophilic species (e.g. $Sn_2S_6^{4-}$, SCN^-, S^{2-}) that offer electrostatic stabilization of nanoparticles in polar media and narrow interparticle separation in nanoparticle solids.[188–190] Unfortunately, in contrast to nanoparticles stabilized by (non-conductive) charged aliphatic ligands, experimental conditions for reliable assembly of nanoparticles stabilized with such compact, charged ligands have yet to be established. Current investigations are restricted to films of short-range ordered particles. It was observed that this new class of surface ligands "silences" grain boundaries in nanostructured solids and enables mobility of charge carriers approaching that of single crystals.[191] Nevertheless, self-assembly of nanoparticles with novel ligand chemistries into ordered arrays remains the ultimate goal. Currently available superlattices have structural disorder that results in low-coordination sites that electronically couple weakly with their nearest neighbors.

A recent development is the creation of epitaxially connected quantum dot films that are obtained by removing the native ligands to induce neck formation (i.e. atomic connections) of specific nanoparticle facets.[192,193] The generated percolating network of quantum dots connected by atomic bonds has an electronic structure that is distinct from that of arrays of quantum dots coupled by ligand molecules.[194] Theoretical analysis in this work concludes that charges are coherently delocalized along segments with a length near 30 nm and charge transport occurs by incoherent charge transfer steps. The localization of electrons at defects has to be considered,[195] and future work requires overcoming limitations, possibly through more uniform epitaxial connections to improve the properties of these quasi-2D quantum dot solids.

1.7.2 *Luminescence*

Self-assembly offers the ability to orient nanoparticles along a common crystallographic axis, a feature that may be exploited in device applications. For example,

CdSe nanoparticles experience a sharp transition from plane polarized to linearly polarized emission, when the nanoparticles are slightly elongated.[196] Assembly of such nanoparticles into smectic superlattice films can be directed by shape anisotropy,[145] by solvent flow or external fields,[118] alignment *via* direct mechanical rubbing of nanoparticle layer,[197] dispersal in stretched polymer matrix,[198] and nematic liquid crystal solvents.[199] Arrays of unidirectionally aligned nanorods are materials with strong polarization anisotropy across cm^2-sized areas. Long-range orientational registry of nanoparticles with linearly polarized emission is useful for improving the efficiency of liquid crystal displays (LCDs).[200] Current technology relies on filtering light from an unpolarized source, resulting in half the photons rejected by the first polarizer in the LCD stack. Orientationally aligning luminescent nanorods can increase the fraction of backlight photons reaching the viewer by absorbing unpolarized light and reemitting polarized light.

1.7.3 *Sensing*

The collective oscillation of free electrons, typically in noble metal nanoparticle superlattices, the so-called plasmon resonance, provides a feasible way to realize light concentration and manipulation on a small scale.[201] The coupling of electromagnetic fields with a superlattice is mediated by surface plasmons, depends on the shape and arrangement of the nanoparticles,[202] and is sensitive to the presence of local chirality in assemblies,[203] as found for example in films formed by Au nanorod template by cholesteric (chiral) liquid crystalline cellulose nanoparticles.[204,205] In another example, achiral Au/Ag core/shell nanocubes act as plasmonic reporters of chirality for attached DNA molecules by providing two orders of magnitude circular dichroism enhancement in the near-visible region.[206] Furthermore, superparticles of polystyrene-stabilized noble metal nanoparticles can have optical properties that are highly sensitive or remarkably independent of cluster orientation, depending on particle number and cluster geometry.[207]

Superlattices of Au nanorods with plasmonic antennae enhancement of the electrical field can detect prions in biological media such as serum and blood (Figure 1.13).[208] Similarly, in BNSLs the strength of near-field plasmonic coupling can be engineered by varying the nanoparticle size, composition, and the lattice symmetry of BNSLs, leading to broadband spectral tunability of the collective plasmonic response of BNSLs across the entire visible spectrum.[209] An application of the plasmon resonance is the SERS effect, which allows sensitive detection of trace chemical or biological species and in Au nanostructures again depends on the nanoparticle core shape.[210–212] Recent work included nanoparticle–polymer composites and expands the study of plasmonic coupling to semiconductor disks.[213]

DNA-based nanostructures are promising candidates for developing self-assembled plasmonic materials.[214] Recently, plasmonic photonic crystals have

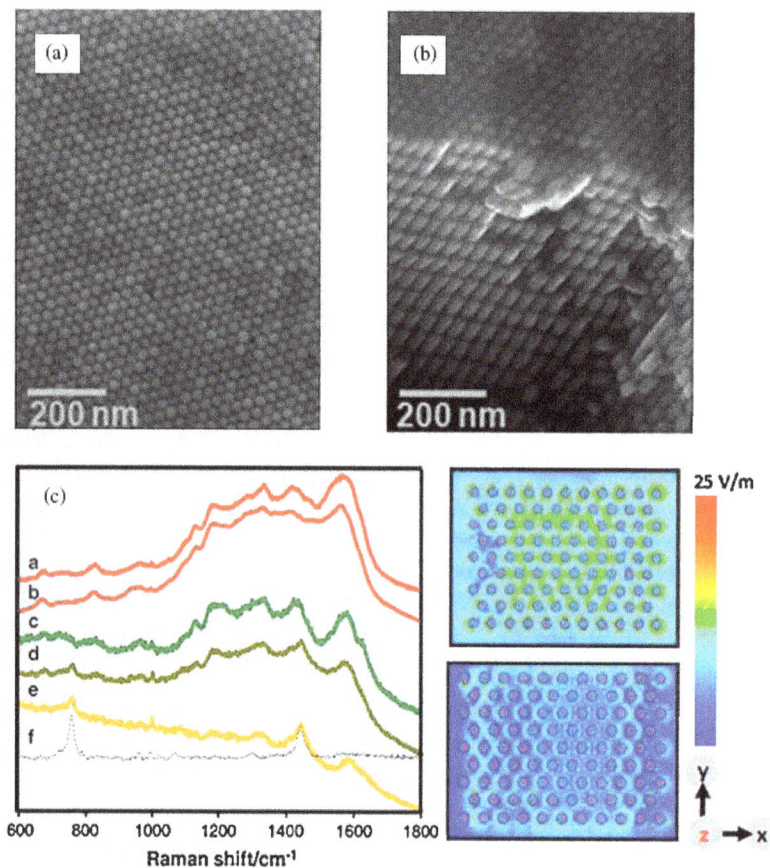

Figure 1.13   SEM images of a typical nanorod supercrystal island film: (a) top view and (b) edge; (c) Left: Surface Enhanced Raman Spectroscopy (SERS) spectra of (a) natural and (b) spiked human blood; (c) natural and (d) spiked human plasma. (e) SERS spectra spiked human plasma after spectral subtraction of the matrix (human plasma). (f) SERS spectra of the scrambled prion; electric field enhancement maps calculated for the top part of a three layer rod-stacked supercrystal (upper right) and for the same location in a single monolayer (down right). Adapted with permission from Ref. 208. Copyright 2011 National Academy of Sciences.

been explored as a way to couple light and plasmon interations by independently adjusting lattice constants and Au nanoparticle diameters, as well as crystal habit (external shape).[215,216] Two distinct optical modes can be controlled independently: a gap mode between the particle and the surface and a lattice mode that originates from cooperative scattering of many nanoparticles in the array.[217]

1.7.4 *Synergistic Effects of Multicomponent Assembled Structures*

Colloidal nanoparticles have chemical and physical functionality of the inorganic core screened from the external environment by a passivating layer of surface ligands. Co-crystallization into a variety of superlattices with precise interparticle separation, stoichiometry, and lattice symmetry integrates this functionality and offers a compelling platform to explore interactions between chemically distinct nanoparticles.

One important class of multifunctional nanomaterials is the semiconductor–metal system. Interaction between luminescent semiconductor nanoparticles and plasmons from nearby metal nanostructures can lead to strong fluorescence enhancement or quenching of emission depending on the spatial arrangement of the components, as well as the energies of exciton and plasmon modes. For example, placement of luminescent CdSe/CdS nanoparticles within a polymer matrix atop a metallic film with nanoscale roughness increases the nanoparticle fluorescence intensity 30-fold when plasmonic and luminescent components are separated by about 35 nm.[218] On the other hand, close-packing of semiconductor and metal nano-particles into BNSLs results in much narrower separations on the order of 2 nm. Optical studies of AlB_2 BNSLs comprised of CdSe and Au nanoparticles revealed a fluorescence intensity of just 5–15% of that of pure CdSe superlattices when a simi-lar number of quantum dots is excited. Time-resolved emission transients show a considerably shorter emission lifetime for CdSe nanoparticles embedded in metal-containing arrays, suggesting creation of an additional non-radiative decay channel *via* energy transfer to the surrounding Au nanoparticles in the binary array.

Self-assembly of magnetic nanoparticles into ordered superlattices, with indi-vidual nanoparticles stable against spontaneous magnetization reversal at room temperature, is a promising approach to increase the density of magnetic data storage. Naturally, understanding the interactions between close-packed magnetic particles is important to the development of this field. Along these lines, co-crystallization of two sizes of magnetic nanoparticles has been shown to yield BNSLs with single-phase-like magnetization alignment and shows one phase magnetization switching behavior and magnetoresistance switching behavior below 200 K.[219,220] Although separate arrays of the two nanoparticle components have different blocking temperatures, determined by the maximum in zero-field-cooled magnetization curves, the BNSLs show zero-field cooled curves with just a single peak. Such an effect is attributed to the collective dipolar interaction between nanoparticle components, suggesting magnetic moments of the smaller nanoparticles are pinned to those of the larger nanoparticle sublattice. Along these lines, BNSLs represent a promising route to produce exchange-spring magnets, nanocomposites in which grains of magnetically hard (high-coercivity) materials are embedded in a magnetically soft (low-coercivity) matrix. For example, co-crystallization of FePt and Fe_3O_4 nanoparticles and thermal annealing to remove organic ligands produced a binary solid of hard FePt and soft Fe_3Pt with significantly higher magnetic performance than either material on its own.[221]

Compositional mixing by co-crystallization of various nanocomponents also presents unique possibilities for tuning the electronic properties of nanoparticle solids. Generally, superlattices produced by self-assembly of nanoparticles capped with hydrocarbon surface ligands possess very low conductivity (G <10^{-11} S/cm). Chemical activation of nanoparticle solids by treatment with small molecules that displace insulating native ligands and reduce interparticle spacing (e.g. hydrazine, thiocyanate) is required to increase conductivity. In some cases, such treatment is sufficiently gentle to preserve the nanoparticle superlattice packing arrangement. However, significant volume contraction and concomitant cracking of nanoparticle films are generally unavoidable, though they may be partially addressed through repeated deposition steps.

The structure and composition of nanoparticle superlattices appear to strongly affect electronic properties such as conductivity by altering the number of charge carriers and/or carrier mobility through the solid. For example, following film exposure to hydrazine and mild thermal annealing at 150°C, the conductivity of binary superlattices comprised of PbTe and Ag_2Te nanoparticles was found to be up to a factor of 10^3 larger than that of corresponding single-component films treated the same way.[222] It was suggested that intimate intermixing of the distinct nanoparticle components may give rise to surface transfer p-type doping. Structure dependent conductivity has also been observed in thermally annealed BNSL membranes containing Fe_3O_4 and FePt nanoparticles across AlB_2 and $NaZn_{13}$ packing arrangements.[223] Compositionally mixed, ordered nanoparticle solids also provide a platform for tuning conductivity *via* establishment of percolation networks through a host lattice. This concept was demonstrated using PbSe nanoparticle superlattices containing metal nanoparticles in random positions as substitutional point defects within the PbSe matrix.[224] Following solid-state thiocyanate film treatment, the conductivity was modulated over six orders of magnitude, from 10^{-6} S/cm for pure PbSe films to 10 S/cm for PbSe containing about 17% metal nanoparticles, the number of nanoparticles required to establish long-range connectivity between randomly distributed metal nanoparticles.

Heterogeneous catalysis is another area that can greatly benefit from self-assembly of compositionally heterogeneous nanoparticle superlattices.[225] Contact with a metal oxide surface, for example, makes otherwise-inert Au nanoparticles highly catalytically active for the oxidation of carbon monoxide.[226] Binary superlattices with a high density of interfacial contact area offer an attractive platform for engineering such catalytic contacts, as demonstrated using Pd–Pt and Au–Fe_3O_4 material pairs.[227,228] In the first example, hydrocarbon-capped octahedral Pt nanoparticles, terminated by catalytically active Pt (111) facets, were co-crystallized with spherical Pd nanoparticles into binary superlattices. Treating these Pt–Pd binary superlattices with UV/ozone and thermal annealing at 180°C removed the surface ligands and established contact between metal nanoparticle surfaces, yielding a nanostructured solid capable of catalyzing the reduction of oxygen to water several times faster than the largely phase separated random mixture or sample with no Pd nanoparticles present. In the second case, Au nanoparticles were co-crystallized with Fe_3O_4 nanoparticles to yield binary superlattices that are capable of catalyzing oxidation of carbon monoxide. Importantly, the superlattice structural framework provides spatial separation of Au nanoparticles by the more thermally stable metal oxide nanoparticles. For example, while heating to 200°C is sufficient to initiate sintering, coalescence and loss of catalytic activity occur for close-packed Au nanoparticles or disordered binary solids. In contrast, when packed between Fe_3O_4 nanoparticles in binary superlattices, the size of Au nanoparticles is preserved up to 400°C, a sufficient operating temperature for most industrial catalytic processes. Yet another example is the co-assembly of Pt and CeO_2 nanocubes on a silica substrate to catalyze methanol decomposition to produce CO and H_2 on the CeO_2–Pt interface, which is subsequently used for ethylene hydroformylation catalysis by the nearby Pt–SiO_2 interface.[229] Furthermore, binary superlattices offer structural control over sintering, allowing coalescence within, but not across, clusters.

Self-assembly has been proposed as a method to achieve Pt-around-Au nanocomposite[230] or to purify Pt nanoparticles by colloidal recrystallization for the isolation of high quality nanoparticles with narrow size and shape distributions.[231] Catalytic activity can be enhanced significantly by eroding the interior of Pt_3Ni polyhedra, partially in a superlattice, leaving only open nanoframes behind.[232]

1.7.5 *Transport and Mechanical Properties*

The collective mechanical response of ordered nanoparticle solids is comparable to nanoparticle-filled polymers or strong plastics. Ag nanoparticles self-organized into an fcc superlattice can couple mechanically to their neighbors. If excited, they vibrate coherently.[233] While in the case of weak bonding, the ligands become the bottleneck for phonon transport, strong bonding can transport distortion of the superlattice more efficiently. Furthermore, the film thickness itself can quantize phonons as observed with picosecond acoustic techniques.[234]

Atomic force microscopy (AFM) nanoindentation has been a useful technique to probe the elasticity and hardness of nanoparticle superlattices. Such experiments revealed that, for close-packed superlattices comprised of oleic acid- capped PbS nanoparticles, elastic modulus and hardness both increase with increasing size of the inorganic core. Similar changes to superlattice elasticity and hardness result when the organic volume fraction is reduced by exchanging oleic acid for shorter ligands.[235] However, this softening effect of surface ligands does not apply to chains sufficiently long to impart mechanical rigidity through backbone entanglement. For example, entanglement has been implicated in the increase of Young's modulus and fracture toughness with increasing chain length of polymers grafted to nanoparticle surfaces.[236] In addition, crystallographic defects and unbound ligands present within the superlattice can soften such structures.[237]

Molecular dynamics simulations also stress the importance of the ligand end group for a macroscopic mechanical response.[238] When long-chain dithiol ligands and spherical aggregates instead of individual nanoparticles were used for self-assembly, then the nanocomposite became plastic and moldable against arbitrarily shaped masters.[239] Thermal hardening produced polycrystalline metal structures of controllable porosity.

Unusual mechanical behavior of highly ordered monolayers of dodecanethiol-ligated Au nanoparticles has been demonstrated by Jaeger and co-workers.[240] Free-standing sheets are flexible enough to bend while draping over edges[241] and can be excited by vibration in the megahertz range into well-defined eigenmodes like drumhead resonators.[242] Au nanoparticle monolayers curled up into hollow scrolls have been indented by atomic force microscopy to extract bending and stretching moduli.[243] Although quite flexible, because they are so thin, the monolayers' resistance to bending is much larger than standard elasticity would predict based on the in-plane stretching behavior. When attached to elastomeric substrates, the same monolayers and multilayers subjected to tensile stress can be forced to fracture.[244] With increasing particle size, the fracture strength increases, while it decreases with film thickness.

Self-assembled membranes also demonstrate selective permeability (permselectivity), which has potential for applications such as ultra- and nanofiltration and chem-selective separation.[245] Molecular transport through Au nanoparticle membranes is two orders of magnitude faster than through polymer-based filters.[246] Au nanoparticles can control ion transport through porous polycarbonate substrates by affecting the substrate electrostatic interactions with ions at the pores.[247] Finally, nanoparticle membranes are interesting for high performance lithium ion battery anodes.[248]

1.8 Conclusion and Outlook

Driven by the continuous demand of size reduction of modern devices, a paradigm shift emerges in production technologies from traditional pick-and-place of individual components to assembly methods based on self-organization concepts prevalent in nature. Since Bentzon and colleagues published electron microscopy images of small ordered arrays of iron oxide nanoparticles in 1989, the library of nanocrystal shapes, sizes, and chemical compositions expanded dramatically. New experimental techniques to produce, characterize, and manipulate nanocrystal assemblies have unlocked a stunningly diverse set of new structures. Beyond the basic evaporation and destabilization approaches to superlattice preparation, new techniques such as self-assembly at immiscible liquid interfaces have opened up opportunities to prepare single-crystalline superlattices across cm^2 areas.

A combination of forces acting between colloidal nanocrystals can generate complex and counterintuitive nanocrystal assemblies. We have highlighted how simple spherical building blocks are used to create assembly structures with properties not inherent to the individual particles. The use of more complex building blocks with non-trivial shapes and made from functional materials will enable access to particle lattices with even more sophisticated symmetries and properties arising from the combination of the inherent material properties themselves with structural features over multiple length scales. Beyond the structure and chemical composition of the nanocrystal building blocks, external factors (e.g. temperature, solvent–ligand interaction, and meniscus dynamics at the drying front) can play an important role in determining which superstructures are formed.

To carry nanocrystal self-assembly from an intellectual curiosity to a practical toolset, researchers in the field must fully understand and elaborate its competitive advantages over other materials systems. For example, self-assembled multicomponent nanocrystal superlattices present a convenient platform to produce precisely intermixed inorganic materials of technologically relevant compositions, offering opportunities to engineer multicomponent heterogeneous catalysts and electronic heterojunctions at the nanoscale. Nanocrystal assemblies offer arrangement of functional inorganic materials (metals, semiconductors, etc.) with 3D precision that exceeds the resolution of current photolithography and imprint lithography. In this aspect, this chapter provides a comprehensive overview on the technological developments for hierarchical nano- and microstructure formation in 2D and 3D. It is the great strength and beauty of colloidal self-assembly to create complex materials with structural hierarchies covering the mesoscopic realm by comparably simple and fast processes, while utilizing cheap starting materials and simple building blocks.

It has recently been recognized that colloidal nanocrystals represent an important class of materials for solution-processed electronic and optoelectronic devices. For instance, arrays of colloidal semiconductor nanocrystals have been utilized in field effect-transistor channels as well as absorber layers of solar cells and photodetectors. Similarly, assemblies of metallic and magnetic nanocrystals can be utilized as solution-processed elements of plasmonic circuits and magnetic inductors. In all these applications, the properties of individual nanocrystals should be complemented by suitable properties of the interparticle medium. It has been shown in numerous studies that charge and heat transport in nanocrystal arrays is very sensitive to the interparticle medium. Along these lines, the development of electrically conductive inorganic ligands resulted in several breakthroughs in nanocrystal electronics and photovoltaics. At the same time, realization of robust self-assembly methods for colloidal nanocrystals with electrically conductive ligands is an important and currently unresolved challenge. Mastering this approach may revolutionize the engineering of epitaxial nanoheterostructures for high-performance applications.

The future for nanocrystal assemblies looks diverse and bright. Continued development of non-traditional surface chemistries, techniques for superlattice preparation, rationalization of phase behavior, and control of the assembly environment will help to bridge fundamental research and practical applications, opening the door to a new generation of functional materials.

References

1. Peng X. G. Manna L., Yang W. D. *et al.* (2000). Shape control of CdSe nanocrystals, *Nature*, 404, 59–61.
2. Cozzoli P. D., Pellegrino T. and Manna L. (2006). Synthesis, properties and perspectives of hybrid nanocrystal structures, *Chem. Soc. Rev.*, 35, 1195–1208.
3. Lu Z. and Yin Y. (2012). Colloidal nanoparticle clusters: Functional materials by design, *Chem. Soc. Rev.*, 41, 6874–6887.
4. Zhuang Z., Peng Q. and Li Y. (2011). Controlled synthesis of semiconductor nanostructures in the liquid phase, *Chem. Soc. Rev.*, 40, 5492–5513.
5. Klajn R., Bishop K. J. M. and Grzybowski B. A. (2007). Light-controlled self-assembly of reversible and irreversible nanoparticle suprastructures, *Proc. Natl. Acad. Sci. USA*, 104, 10305–10309.
6. Sun S. H., Murray C. B., Weller D., Folks L. and Moser A. (2000). Monodisperse fept nanoparticles and ferromagnetic fept nanocrystal superlattices, *Science*, 287, 1989–1992.
7. Nie Z., Petukhova A. and Kumacheva E. (2010). Properties and emerging applications of self-assembled structures made from inorganic nanoparticles, *Nat. Nanotechnol.*, 5, 15–25.
8. Schall P., Cohen I., Weitz D. A. and Spaepen F. (2004). Visualization of dislocation dynamics in colloidal crystals, *Science*, 305, 1944–1948.
9. Schall P., Cohen I., Weitz D. A. and Spaepen F. (2006). Visualizing dislocation nucleation by indenting colloidal crystals, *Nature*, 440, 319–323.
10. Kozina A., Sagawe D., Diaz-Leyva P., Bartsch E. and Palberg T. (2012). Polymer-enforced crystallization of a eutectic binary hard sphere mixture, *Soft Matter*, 8, 627–630.
11. Bausch A. R., Bowick M. J., Cacciuto A. *et al.* (2003). Grain boundary scars and spherical crystallography, *Science*, 299, 1716–1718.

12. Meng G., Paulose J., Nelson D. R. and Manoharan V. N. (2014). Elastic instability of a crystal growing on a curved surface, *Science*, 343, 634–637.
13. Raccis R., Nikoubashman A., Retsch M. *et al.* (2011). Confined diffusion in periodic porous nanostructures, *ACS Nano*, 5, 4607–4616.
14. Yin J., Retsch M., Thomas E. L. and Boyce M. C. (2012). Collective mechanical behavior of multilayer colloidal arrays of hollow nanoparticles, *Langmuir*, 28, 5580–5588.
15. Rao K. D. M., Hunger C., Gupta R., Kulkarni G. U. and Thelakkat M. (2014). A cracked polymer templated metal network as a transparent conducting electrode for ito-free organic solar cells, *Phys. Chem. Chem. Phys.*, 16, 15107–15110.
16. Lee J.-H., Singer J. P. and Thomas E. L. (2012). Micro-/nanostructured mechanical metamaterials, *Adv. Mater.*, 24, 4782–4810.
17. Galisteo-Lopez J. F., Ibisate M., Sapienza R. *et al.* (2011). Self-assembled photonic structures, *Adv. Mater.*, 23, 30–69.
18. Cheng W., Wang J., Jonas U., Fytas G. and Stefanou N. (2006). Observation and tuning of hypersonic bandgaps in colloidal crystals, *Nat. Mater.*, 5, 830–836.
19. Pikul J. H., Zhang H. G., Cho J., Braun P. V. and King W. P. (2013). High-power lithium ion microbatteries from interdigitated three-dimensional bicontinuous nanoporous electrodes, *Nat. Commun.*, 4.
20. Tetreault N., Arsenault E., Heiniger L.-P. *et al.* (2011). High-efficiency dye-sensitized solar cell with three-dimensional photoanode, *Nano Lett.*, 11, 4579–4584.
21. Ruckdeschel P., Kemnitzer T. W., Nutz F. A., Senker J. and Retsch M. (2015). Hollow silica sphere colloidal crystals: Insights into calcination dependent thermal transport, *Nanoscale*, 7, 10059–10070.
22. Fenzl C., Hirsch T. and Wolfbeis O. S. (2014). Photonic crystals for chemical sensing and biosensing, *Angew. Chem. Int. Ed. Engl.*, 53, 3318–3335.
23. Anker J. N., Hall W. P., Lyandres O. *et al.* (2008). Biosensing with plasmonic nanosensors, *Nat. Mater.*, 7, 442–453.
24. Lattuada M. and Hatton T. A. (2011). Synthesis, properties and applications of janus nanoparticles, *Nano Today*, 6, 286–308.
25. Niu W. and Xu G. (2011). Crystallographic control of noble metal nanocrystals, *Nano Today*, 6, 265–285.
26. Grzelczak M., Vermant J., Furst E. M. and Liz-Marzan L. M. (2010). Directed self-assembly of nanoparticles, *ACS Nano*, 4, 3591–3605.
27. Vanmaekelbergh D. (2011). Self-assembly of colloidal nanocrystals as route to novel classes of nanostructured materials, *Nano Today*, 6, 419–437.
28. Ye X., Jin L., Caglayan H. *et al.* (2012). Improved size-tunable synthesis of monodisperse gold nanorods through the use of aromatic additives, *ACS Nano*, 6, 2804–2817.
29. Wang X., Zhuang J., Peng Q. and Li Y. D. (2005). A general strategy for nanocrystal synthesis, *Nature*, 437, 121–124.

30. Yin Y. and Alivisatos A. P. (2005). Colloidal nanocrystal synthesis and the organic-inorganic interface, *Nature*, 437, 664–670.
31. Pileni M. P. (2007). Self-assembly of inorganic nanocrystals: Fabrication and collective intrinsic properties, *Acc. Chem. Res.*, 40, 685–693.
32. Talapin D. V., Nelson J. H., Shevchenko E. V. *et al.* (2007). Seeded growth of highly luminescent CdSe/CdS nanoheterostructures with rod and tetrapod morphologies, *Nano Lett.*, 7, 2951–2959.
33. Auyeung E., Cutler J. I., Macfarlane R. J. *et al.* (2012). Synthetically programmable nanoparticle superlattices using a hollow three-dimensional spacer approach, *Nat. Nanotechnol.*, 7, 24–28.
34. Chen Q., Bae S. C. and Granick S. (2011). Directed self-assembly of a colloidal kagome lattice, *Nature*, 469, 381–384.
35. Romano F. and Sciortino F. (2012). Patterning symmetry in the rational design of colloidal crystals, *Nat. Commun.*, 3.
36. Wang Y., Wang Y., Breed D. R. *et al.* (2012). Colloids with valence and specific directional bonding, *Nature*, 491, U51–U61.
37. Groeschel A. H., Walther A., Loebling T. I. *et al.* (2013). Guided hierarchical co-assembly of soft patchy nanoparticles, *Nature*, 503, 247–+.
38. Vogel N., Retsch M., Fustin C.-A., Del Campo A. and Jonas U. (2015). Advances in colloidal assembly: The design of structure and hierarchy in two and three dimensions, *Chem. Rev.*, 115, 6265–6311.
39. Li Q., Jonas U., Zhao X. S. and Kapp M. (2008). The forces at work in collodial self-assembly: A review on fundamental interactions between collodial particels, *Asia Pac. J. Chem. Eng.*, 3, 255–268.
40. Bishop K. J. M., Wilmer C. E., Soh S. and Grzybowski B. A. (2009). Nanoscale forces and their uses in self-assembly, *Small*, 5, 1600–1630.
41. Luo D., Yan C. and Wang T. (2015). Interparticle forces underlying nanoparticle self-assemblies, *Small*, 11, 5984–6008.
42. Pieranski P. (1980). Two-dimensional interfacial colloidal crystals, *Phys. Rev. Lett.*, 45, 569–572.
43. Law A. D., Auriol M., Smith D., Horozov T. S. and Buzza D. M. A. (2013). Self-assembly of two-dimensional colloidal clusters by tuning the hydrophobicity, composition, and packing geometry, *Phys. Rev. Lett.*, 110.
44. Tabor R. F., Grieser F., Dagastine R. R. and Chan D. Y. C. (2014). The hydrophobic force: Measurements and methods, *Phys. Chem. Chem. Phys.*, 16, 18065–18075.
45. Romero-Cano M. S., Martin-Rodriguez A. and De Las Nieves F. J. (2001). Electrosteric stabilization of polymer colloids with different functionality, *Langmuir*, 17, 3505–3511.
46. Dickinson E. (2010). Food emulsions and foams: Stabilization by particles, *Curr. Opin. Colloid Interface Sci.*, 15, 40–49.
47. Boeker A., He J., Emrick T. and Russell T. P. (2007). Self-assembly of nanoparticles at interfaces, *Soft Matter*, 3, 1231–1248.

48. Grzelczak M., Vermant J., Furst E. M. and Liz-Marzan L. M. (2010). Directed self-assembly of nanoparticles, *ACS Nano*, 4, 3591–3605.

49. Kralchevsky P. A. and Nagayama K. (2000). Capillary interactions between particles bound to interfaces, liquid films and biomembranes, *Adv. Colloid Interface Sci.*, 85, 145–192.

50. Weekes S. M., Ogrin F. Y., Murray W. A. and Keatley P. S. (2007). Macroscopic arrays of magnetic nanostructures from self-assembled nanosphere templates, *Langmuir*, 23, 1057–1060.

51. Baranov D., Fiore A., Van Huis M. *et al.* (2010). Assembly of colloidal semiconductor nanorods in solution by depletion attraction, *Nano Lett.*, 10, 743–749.

52. Kraft D. J., Ni R., Smallenburg F. *et al.* (2012). Surface roughness directed self-assembly of patchy particles into colloidal micelles, *Proc. Natl. Acad. Sci. U. S. A.*, 109, 10787–10792.

53. Murray C. B., Kagan C. R. and Bawendi M. G. (1995). Self-organization of CdSe nanocrystallites into 3-dimensional quantum-dot superlattices, *Science*, 270, 1335–1338.

54. Jana N. R. (2004). Shape effect in nanoparticle self-assembly, *Angew. Chem. Int. Ed. Engl.*, 43, 1536–1540.

55. Sau T. K. and Murphy C. J. (2005). Self-assembly patterns formed upon solvent evaporation of aqueous cetyltrimethylammonium bromide-coated gold nanoparticles of various shapes, *Langmuir*, 21, 2923–2929.

56. Singh A., Gunning R. D., Ahmed S. *et al.* (2012). Controlled semiconductor nanorod assembly from solution: Influence of concentration, charge and solvent nature, *J. Mater. Chem.*, 22, 1562–1569.

57. Kinge S., Crego-Calama M. and Reinhoudt D. N. (2008). Self-assembling nanoparticles at surfaces and interfaces, *ChemPhysChem.*, 9, 20–42.

58. Min Y., Akbulut M., Kristiansen K., Golan Y. and Israelachvili J. (2008). The role of interparticle and external forces in nanoparticle assembly, *Nat. Mater.*, 7, 527–538.

59. Malaquin L., Kraus T., Schmid H., Delamarche E. and Wolf H. (2007). Controlled particle placement through convective and capillary assembly, *Langmuir*, 23, 11513–11521.

60. Born P., Blum S., Munoz A. and Kraus T. (2011). Role of the meniscus shape in large-area convective particle assembly, *Langmuir*, 27, 8621–8633.

61. Still T., Yunker P. J. and Yodh A. G. (2012). Surfactant-induced marangoni eddies alter the coffee-rings of evaporating colloidal drops, *Langmuir*, 28, 4984–4988.

62. Cui L., Zhang J., Zhang X. *et al.* (2012). Suppression of the coffee ring effect by hydrosoluble polymer additives, *ACS Appl. Mater. Interfaces*, 4, 2775–2780.

63. Yunker P. J., Still T., Lohr M. A. and Yodh A. G. (2011). Suppression of the coffee-ring effect by shape-dependent capillary interactions, *Nature*, 476, 308–311.

64. Armstrong E., Khunsin W., Osiak M. *et al.* (2014). Ordered 2d colloidal photonic crystals on gold substrates by surfactant-assisted fast-rate dip coating, *Small*, 10, 1895–1901.

65. Born P., Munoz A., Cavelius C. and Kraus T. (2012). Crystallization mechanisms in convective particle assembly, *Langmuir*, 28, 8300–8308.

66. Ming T., Kou X., Chen H. *et al.* (2008). Ordered gold nanostructure assemblies formed by droplet evaporation, *Angew. Chem. Int. Ed. Engl.*, 47, 9685–9690.

67. Gelbart W. M., Sear R. P., Heath J. R. and Chaney S. (1999). Array formation in nano-colloids: Theory and experiment in 2D, *Faraday Discuss.*, 112, 299–307.

68. Rabani E., Reichman D. R., Geissler P. L. and Brus L. E. (2003). Drying-mediated self-assembly of nanoparticles, *Nature*, 426, 271–274.

69. Kang C.-C., Lai C.-W., Peng H.-C., Shyue J.-J. and Chou P.-T. (2008). 2d self-bundled cds nanorods with micrometer dimension in the absence of an external directing process, *ACS Nano*, 2, 750–756.

70. Rupich S. M., Shevchenko E. V., Bodnarchuk M. I., Lee B. and Talapin D. V. (2010). Size-dependent multiple twinning in nanocrystal superlattices, *J. Am. Chem. Soc.*, 132, 289–296.

71. Talapin D. V., Shevchenko E. V., Murray C. B. *et al.* (2004). CdSe and CdSe/CdS nanorod solids, *J. Am. Chem. Soc.*, 126, 12984–12988.

72. Abecassis B., Tessier M. D., Davidson P. and Dubertret B. (2014). Self-assembly of CdSe nanoplatelets into giant micrometer-scale needles emitting polarized light, *Nano Lett.*, 14, 710–715.

73. Zhuang J., Wu H., Yang Y. and Cao Y. C. (2007). Supercrystalline colloidal particles from artificial atoms, *J. Am. Chem. Soc.*, 129, 14166–+.

74. Wang T., Zhuang J., Lynch J. *et al.* (2012). Self-assembled colloidal superparticles from nanorods, *Science*, 338, 358–363.

75. Onsager L. (1949). The effects of shape on the interaction of colloidal particles, *Ann. N.Y. Acad. Sci.*, 51, 627–659.

76. Park K., Koerner H. and Vaia R. A. (2010). Depletion-induced shape and size selection of gold nanoparticles, *Nano Lett.*, 10, 1433–1439.

77. Young K. L., Jones M. R., Zhang J. *et al.* (2012). Assembly of reconfigurable one-dimensional colloidal superlattices due to a synergy of fundamental nanoscale forces, *Proc. Natl. Acad. Sci. USA*, 109, 2240–2245.

78. Young K. L., Personick M. L., Engel M. *et al.* (2013). A directional entropic force approach to assemble anisotropic nanoparticles into superlattices, *Angew. Chem. Int. Ed. Engl.*, 52, 13980–13984.

79. Zhang Y., Lu F., Van Der Lelie D. and Gang O. (2011). Continuous phase transformation in nanocube assemblies, *Phys. Rev. Lett.*, 107.

80. Li R., Bian K., Wang Y. *et al.* (2015). An obtuse rhombohedral superlattice assembled by pt nanocubes, *Nano Lett.*, 15, 6254–6260.

81. Karas A. S., Glaser J. and Glotzer S. C. (2016). Using depletion to control colloidal crystal assemblies of hard cuboctahedra, *Soft Matter*, 12, 5199–5204.

82. Rossi L., Soni V., Ashton D. J. *et al.* (2015). Shape-sensitive crystallization in colloidal superball fluids, *Proc. Natl. Acad. Sci. USA*, 112, 5286–5290.

83. Rossi L., Sacanna S., Irvine W. T. M. *et al.* (2011). Cubic crystals from cubic colloids, *Soft Matter*, 7, 4139–4142.

84. Mahynski N. A., Panagiotopoulos A. Z., Meng D. and Kumar S. K. (2014). Stabilizing colloidal crystals by leveraging void distributions, *Nat. Commun.*, 5.

85. Mahynski N. A., Rovigatti L., Likos C. N. and Panagiotopoulos A. Z. (2016). Bottom-up colloidal crystal assembly with a twist, *ACS Nano*, 10, 5459–5467.
86. Song R.-Q. and Coelfen H. (2011). Additive controlled crystallization, *Crystengcomm*, 13, 1249–1276.
87. Henzie J., Gruenwald M., Widmer-Cooper A., Geissler P. L. and Yang P. (2012). Self-assembly of uniform polyhedral silver nanocrystals into densest packings and exotic superlattices, *Nat. Mater.*, 11, 131–137.
88. Bodnarchuk M. I., Kovalenko M. V., Heiss W. and Talapin D. V. (2010). Energetic and entropic contributions to self-assembly of binary nanocrystal superlattices: Temperature as the structure-directing factor, *J. Am. Chem. Soc.*, 132, 11967–11977.
89. Cui Y., Bjork M. T., Liddle J. A. *et al.* (2004). Integration of colloidal nanocrystals into lithographically patterned devices, *Nano Lett.*, 4, 1093–1098.
90. Jiang L., Chen X., Lu N. and Chi L. (2014). Spatially confined assembly of nanoparticles, *Acc. Chem. Res.*, 47, 3009–3017.
91. Zhou Y., Zhou X., Park D. J. *et al.* (2014). Shape-selective deposition and assembly of anisotropic nanoparticles, *Nano Lett.*, 14, 2157–2161.
92. Fan J. A., Bao K., Sun L. *et al.* (2012). Plasmonic mode engineering with templated self-assembled nanoclusters, *Nano Lett.*, 12, 5318–5324.
93. Greybush N. J., Saboktakin M., Ye X. *et al.* (2014). Plasmon-enhanced upconversion luminescence in single nanophosphor-nanorod heterodimers formed through template-assisted self-assembly, *ACS Nano*, 8, 9482–9491.
94. Rupich S. M., Castro F. C., Irvine W. T. M. and Talapin D. V. (2014). Soft epitaxy of nanocrystal superlattices, *Nat. Commun.*, 5.
95. Saunders A. E. and Korgel B. A. (2004). Second virial coefficient measurements of dilute gold nanocrystal dispersions using small-angle x-ray scattering, *J. Phys. Chem. B*, 108, 16732–16738.
96. Goodfellow B. W., Rasch M. R., Hessel C. M. *et al.* (2013). Ordered structure rearrangements in heated gold nanocrystal superlattices, *Nano Lett.*, 13, 5710–5714.
97. Wei J., Schaeffer N. and Pileni M.-P. (2016). Solvent-mediated crystallization of nanocrystal 3d assemblies of silver nanocrystals: Unexpected superlattice ripening, *Chem. Mater.*, 28, 293–302.
98. Yu Y., Jain A., Guillaussier A. *et al.* (2015). Nanocrystal superlattices that exhibit improved order on heating: An example of inverse melting?, *Faraday Discuss.*, 181, 181–192.
99. Lewandowski W., Fruhnert M., Mieczkowski J., Rockstuhl C. and Gorecka E. (2015). Dynamically self-assembled silver nanoparticles as a thermally tunable metamaterial, *Nat. Commun.*, 6.
100. Lalatonne Y., Richardi J. and Pileni M. P. (2004). Van der waals versus dipolar forces controlling mesoscopic organizations of magnetic nanocrystals, *Nat. Mater.*, 3, 121–125.
101. Ryan K. M., Mastroianni A., Stancil K. A., Liu H. and Alivisatos A. P. (2006). Electric-field-assisted assembly of perpendicularly oriented nanorod superlattices, *Nano Lett.*, 6, 1479–1482.

102. Yanai N., Sindoro M., Yan J. and Granick S. (2013). Electric field-induced assembly of monodisperse polyhedral metal-organic framework crystals, *J. Am. Chem. Soc.*, 135, 34–37.

103. Yanai N. and Granick S. (2012). Directional self-assembly of a colloidal metal-organic framework, *Angew. Chem. Int. Ed. Engl.*, 51, 5638–5641.

104. Barsotti R. J. Jr. Vahey M. D., Wartena R. *et al.* (2007). Assembly of metal nanoparticles into nanogaps, *Small*, 3, 488–499.

105. Smallenburg F., Vutukuri H. R., Imhof A., Van Blaaderen A. and Dijkstra, M. (2012). Self-assembly of colloidal particles into strings in a homogeneous external electric or magnetic field, *J. Phys.-Condes. Matter*, 24.

106. Ye L., Pearson T., Cordeau Y., Mefford O. T. and Crawford T. M. (2016). Triggered self-assembly of magnetic nanoparticles, *Scientific Reports*, 6.

107. Singh G., Chan H., Baskin A. *et al.* (2014). Self-assembly of magnetite nanocubes into helical superstructures, *Science*, 345, 1149–1153.

108. Srivastava S., Santos A., Critchley K. *et al.* (2010). Light-controlled self-assembly of semiconductor nanoparticles into twisted ribbons, *Science*, 327, 1355–1359.

109. Yeom J., Yeom B., Chan H. *et al.* (2015). Chiral templating of self-assembling nanostructures by circularly polarized light, *Nat. Mater.*, 14, 66–72.

110. Donaldson J. G. and Kantorovich S. S. (2015). Directional self-assembly of permanently magnetised nanocubes in quasi two dimensional layers, *Nanoscale*, 7, 3217–3228.

111. Taheri S. M., Michaelis M., Friedrich T. *et al.* (2015). Self-assembly of smallest magnetic particles, *Proc. Natl. Acad. Sci. U. S. A.*, 112, 14484–14489.

112. Zhang X., Zhang Z. and Glotzer S. C. (2007). Simulation study of dipole-induced self-assembly of nanocubes, *J. Phys. Chem. C*, 111, 4132–4137.

113. Huang S., Pessot G., Cremer P. *et al.* (2016). Buckling of paramagnetic chains in soft gels, *Soft Matter*, 12, 228–237.

114. Kantorovich S. S., Ivanov A. O., Rovigatti L., Tavares J. M. and Sciortino F. (2015). Temperature-induced structural transitions in self-assembling magnetic nanocolloids, *Phys. Chem. Chem. Phys.*, 17, 16601–16608.

115. Esquivel-Sirvent R. and Schatz G. C. (2013). Van der waals torque coupling between slabs composed of planar arrays of nanoparticles, *J. Phys. Chem. C*, 117, 5492–5496.

116. Yasui K. and Kato K. (2015). Oriented attachment of cubic or spherical batio3 nanocrystals by van der waals torque, *J. Phys. Chem. C*, 119, 24597–24605.

117. Kundu P. K., Samanta D., Leizrowice R. *et al.* (2015). Light-controlled self-assembly of non-photoresponsive nanoparticles, *Nat. Chem.*, 7, 646–652.

118. Carbone L., Nobile C., De Giorgi M. *et al.* (2007). Synthesis and micrometer–scale assembly of colloidal CdSe/CdS nanorods prepared by a seeded growth approach, *Nano Lett.*, 7, 2942–2950.

119. Gupta S., Zhang Q., Emrick T. and Russell T. P. (2006). "Self-corralling" nanorods under an applied electric field, *Nano Lett.*, 6, 2066–2069.

120. Hu Z., Fischbein M. D., Querner C. and Drndic M. (2006). Electric-field-driven accumulation and alignment of CdSe and CdTe nanorods in nanoscale devices, *Nano Lett.*, 6, 2585–2591.

121. Smith P. A., Nordquist C. D., Jackson T. N. *et al.* (2000). Electric-field assisted assembly and alignment of metallic nanowires, *Appl. Phys. Lett.*, 77, 1399–1401.

122. Lu Y., Yin Y. D. and Xia Y. N. (2001). Three-dimensional photonic crystals with non-spherical colloids as building blocks, *Adv. Mater.*, 13, 415–420.

123. Lee S. H., Song Y., Hosein I. D. and Liddell C. M. (2009). Magnetically responsive and hollow colloids from nonspherical core-shell particles of peanut-like shape, *J. Mater. Chem.*, 19, 350–355.

124. Ding T., Song K., Clays K. and Tung C.-H. (2009). Fabrication of 3d photonic crystals of ellipsoids: Convective self-assembly in magnetic field, *Adv. Mater.*, 21, 1936–1940.

125. Chen M., Pica T., Jiang Y.-B. *et al.* (2007). Synthesis and self-assembly of fcc phase fept nanorods, *J. Am. Chem. Soc.*, 129, 6348–+.

126. Tanase M., Bauer L. A., Hultgren A. *et al.* (2001). Magnetic alignment of fluorescent nanowires, *Nano Lett.*, 1, 155–158.

127. Peng S., Lee Y., Wang C. *et al.* (2008). A facile synthesis of monodisperse au nanoparticles and their catalysis of co oxidation, *Nano Research*, 1, 229–234.

128. Goodfellow B. W., Yu Y., Bosoy C. A., Smilgies D.-M. and Korgel B. A. (2015). The role of ligand packing frustration in body-centered cubic (bcc). superlattices of colloidal nanocrystals, *J. Phys. Chem. Lett.*, 6, 2406–2412.

129. Courty A., Richardi J., Albouy P.-A. and Pileni M.-P. (2011). How to control the crystalline structure of supracrystals of 5-nm silver nanocrystals, *Chem. Mat.*, 23, 4186–4192.

130. Gasperino D., Meng L., Norris D. J. and Derby J. J. (2008). The role of fluid flow and convective steering during the assembly of colloidal crystals, *J. Cryst. Growth*, 310, 131–139.

131. Norris D. J., Arlinghaus E. G., Meng L. L., Heiny R. and Scriven L. E. (2004). Opaline photonic crystals: How does self-assembly work?, *Adv. Mater.*, 16, 1393–1399.

132. Talapin D. V., Shevchenko E. V., Murray C. B., Titov A. V. and Kral P. (2007). Dipole-dipole interactions in nanoparticle superlattices, *Nano Lett.*, 7, 1213–1219.

133. Portales H., Goubet N., Sirotkin S. *et al.* (2012). Crystallinity segregation upon selective self-assembling of gold colloidal single nanocrystals, *Nano Lett.*, 12, 5292–5298.

134. Li R., Bian K., Hanrath T., Bassett W. A. and Wang Z. (2014). Decoding the super-lattice and interface structure of truncate pbs nanocrystal-assembled supercrystal and associated interaction forces, *J. Am. Chem. Soc.*, 136, 12047–12055.

135. Liu K., Zhao N. N. and Kumacheva E. (2011). Self-assembly of inorganic nanorods, *Chem. Soc. Rev.*, 40, 656–671.

136. Zhang S.-Y., Regulacio M. D. and Han M.-Y. (2014). Self-assembly of colloidal one-dimensional nanocrystals, *Chem. Soc. Rev.*, 43, 2301–2323.

137. Jana N. R., Gearheart L. and Murphy C. J. (2001). Wet chemical synthesis of high aspect ratio cylindrical gold nanorods, *J. Phys. Chem. B*, 105, 4065–4067.

138. Jana N. R., Gearheart L. and Murphy C. J. (2001). Wet chemical synthesis of silver nanorods and nanowires of controllable aspect ratio, *Chem. Commun.*, 617–618.

139. Ye X., Zheng C., Chen J., Gao Y. and Murray C. B. (2013). Using binary surfactant mixtures to simultaneously improve the dimensional tunability and monodispersity in the seeded growth of gold nanorods, *Nano Lett.*, 13, 765–771.

140. Ye X., Gao Y., Chen J. *et al.* (2013). Seeded growth of monodisperse gold nanorods using bromide-free surfactant mixtures, *Nano Lett.*, 13, 2163–2171.

141. Ithurria S., Tessier M. D., Mahler B. *et al.* (2011). Colloidal nanoplatelets with two-dimensional electronic structure, *Nat. Mater.*, 10, 936–941.

142. Ye X., Collins J. E., Kang Y. *et al.* (2010). Morphologically controlled synthesis of colloidal upconversion nanophosphors and their shape-directed self-assembly, *Proc. Natl. Acad. Sci. U. S. A.*, 107, 22430–22435.

143. Li L. S., Walda J., Manna L. and Alivisatos A. P. (2002). Semiconductor nanorod liquid crystals, *Nano Lett.*, 2, 557–560.

144. Chen Q., Cho H., Manthiram K. *et al.* (2015). Interaction potentials of anisotropic nanocrystals from the trajectory sampling of particle motion using *in situ* liquid phase transmission electron microscopy, *Acs Central Science*, 1, 33–39.

145. Diroll B. T., Greybush N. J., Kagan C. R. and Murray C. B. (2015). Smectic nanorod superlattices assembled on liquid subphases: Structure, orientation, defects, and optical polarization, *Chem. Mat.*, 27, 2998–3008.

146. Zhuang J., Shaller A. D., Lynch J. *et al.* (2009). Cylindrical superparticles from semiconductor nanorods, *J. Am. Chem. Soc.*, 131, 6084–+.

147. Huo Z., Tsung C.-K., Huang W. *et al.* (2009). Self-organized ultrathin oxide nano-crystals, *Nano Lett.*, 9, 1260–1264.

148. Rowland C. E., Fedin I., Zhang H. *et al.* (2015). Picosecond energy transfer and multiexciton transfer outpaces auger recombination in binary CdSe nanoplatelet solids, *Nat. Mater.*, 14, 484–489.

149. Millan J. A., Ortiz D., Van Anders G. and Glotzer S. C. (2014). Self-assembly of archimedean tilings with enthalpically and entropically patchy polygons, *ACS Nano*, 8, 2918–2928.

150. Zhao K., Bruinsma R. and Mason T. G. (2011). Entropic crystal-crystal transitions of brownian squares, *Proc. Natl. Acad. Sci. U. S. A.*, 108, 2684–2687.

151. Zhao K., Bruinsma R. and Mason T. G. (2012). Local chiral symmetry breaking in triatic liquid crystals, *Nat. Commun.*, 3.

152. Henzie J., Grunwald M., Widmer-Cooper A., Geissler P. L. and Yang P. D. (2012). Self-assembly of uniform polyhedral silver nanocrystals into densest packings and exotic superlattices, *Nat. Mater.*, 11, 131–137.

153. Zhang H., Yang J., Hanrath T. and Wise F. W. (2014). Sub-10 nm monodisperse pbs cubes by post-synthesis shape engineering, *Phys. Chem. Chem. Phys.*, 16, 14640–14643.

154. Dumestre F., Chaudret B., Amiens C., Renaud P. and Fejes P. (2004). Super-lattices of iron nanocubes synthesized from fe n(sime3)(2)(2), *Science*, 303, 821–823.

155. Ren J. and Tilley, R. D. (2007). Preparation, self-assembly, and mechanistic study of highly monodispersed nanocubes, *J. Am. Chem. Soc.*, 129, 3287–3291.

156. Wang T., Wang X., Lamontagne D. *et al.* (2012). Shape-controlled synthesis of colloidal superparticles from nanocubes, *J. Am. Chem. Soc.*, 134, 18225–18228.

157. Gordon T. R., Paik T., Klein D. R. *et al.* (2013). Shape-dependent plasmonic response and directed self-assembly in a new semiconductor building block, indium-doped cadmium oxide (ico)., *Nano Lett.*, 13, 2857–2863.

158. Javon E., Gaceur M., Dachraoui W. *et al.* (2015). Competing forces in the self-assembly of coupled zno nanopyramids, *ACS Nano*, 9, 3685–3694.

159. Damasceno P. F., Engel M. and Glotzer S. C. (2012). Crystalline assemblies and densest packings of a family of truncated tetrahedra and the role of directional entropic forces, *ACS Nano*, 6, 609–614.

160. Boles M. A. and Talapin D. V. (2014). Self-assembly of tetrahedral CdSe nanocrystals: Effective "patchiness" via anisotropic steric interaction, *J. Am. Chem. Soc.*, 136, 5868–5871.

161. Zhang J., Luo Z., Quan Z. *et al.* (2011). Low packing density self-assembled super-structure of octahedral pt3ni nanocrystals, *Nano Lett.*, 11, 2912–2918.

162. Huang M. H. and Thoka S. (2015). Formation of supercrystals through self-assembly of polyhedral nanocrystals, *Nano Today*, 10, 81–92.

163. Nakagawa Y., Kageyama H., Oaki Y. and Imai H. (2014). Direction control of oriented self-assembly for 1d, 2d, and 3d microarrays of anisotropic rectangular nanoblocks, *J. Am. Chem. Soc.*, 136, 3716–3719.

164. Liao C.-W., Lin Y.-S., Chanda K., Song Y.-F. and Huang M. H. (2013). Formation of diverse supercrystals from self-assembly of a variety of polyhedral gold nanocrystals, *J. Am. Chem. Soc.*, 135, 2684–2693.

165. Tan R., Zhu H., Cao C. and Chen O. (2016). Multi-component superstructures self-assembled from nanocrystal building blocks, *Nanoscale*, 8, 9944–9961.

166. Vogel N., De Viguerie L., Jonas U., Weiss C. K. and Landfester K. (2011). Wafer-scale fabrication of ordered binary colloidal monolayers with adjustable stoichiometries, *Adv. Funct. Mater.*, 21, 3064–3073.

167. Vogel N., Weiss C. K. and Landfester K. (2012). From soft to hard: The generation of functional and complex colloidal monolayers for nanolithography, *Soft Matter*, 8, 4044–4061.

168. Shevchenko E. V., Talapin D. V., Kotov N. A., O'brien S. and Murray C. B. (2006). Structural diversity in binary nanoparticle superlattices, *Nature*, 439, 55–59.

169. Talapin D. V., Shevchenko E. V., Bodnarchuk M. I. *et al.* (2009). Quasicrystalline order in self-assembled binary nanoparticle superlattices, *Nature*, 461, 964–967.

170. Yang Z., Wei J., Bonville P. and Pileni M.-P. (2015). Beyond entropy: Magnetic forces induce formation of quasicrystalline structure in binary nanocrystal superlattices, *J. Am. Chem. Soc.*, 137, 4487–4493.

171. Ye X., Chen J. and Murray C. B. (2011). Polymorphism in self-assembled ab(6). binary nanocrystal superlattices, *J. Am. Chem. Soc.*, 133, 2613–2620.

172. Boneschanscher M. P., Evers W. H., Qi W. *et al.* (2013). Electron tomography resolves a novel crystal structure in a binary nanocrystal superlattice, *Nano Lett.*, 13, 1312–1316.

173. Boles M. A. and Talapin D. V. (2015). Many-body effects in nanocrystal superlattices: Departure from sphere packing explains stability of binary phases, *J. Am. Chem. Soc.*, 137, 4494–4502.

174. Hopkins A. B., Stillinger F. H. and Torquato S. (2012). Densest binary sphere packings, *Phys. Rev. E Stat. Nonlin. Soft Matter Phys.*, 85.

175. Evers W. H., Friedrich H., Filion L., Dijkstra M. and Vanmaekelbergh D. (2009). Observation of a ternary nanocrystal superlattice and its structural characterization by electron tomography, *Angew. Chem. Int. Ed. Engl.*, 48, 9655–9657.

176. Dong A., Ye X., Chen J. and Murray C. B. (2011). Two-dimensional binary and ternary nanocrystal superlattices: The case of monolayers and bilayers, *Nano Lett.*, 11, 1804–1809.

177. Ye X., Zhu C., Ercius P. *et al.* (2015). Structural diversity in binary superlattices self-assembled from polymer-grafted nanocrystals, *Nat. Commun.*, 6.

178. Bodnarchuk M. I., Erni R., Krumeich F. and Kovalenko M. V. (2013). Binary superlattices from colloidal nanocrystals and giant polyoxometalate clusters, *Nano Lett.*, 13, 1699–1705.

179. Travesset A. (2015). Binary nanoparticle superlattices of soft-particle systems, *Proc. Natl. Acad. Sci. U. S. A.*, 112, 9563–9567.

180. Likos C. N. (2001). Effective interactions in soft condensed matter physics, *Phys. Rep -Rev. Sect. Phys. Lett.*, 348, 267–439.

181. Ben-Simon A., Eshet H. and Rabani E. (2013). On the phase behavior of binary mixtures of nanoparticles, *ACS Nano*, 7, 978–986.

182. Leunissen M. E., Christova C. G., Hynninen A. P. *et al.* (2005). Ionic colloidal crystals of oppositely charged particles, *Nature*, 437, 235–240.

183. Kovalenko M. V., Manna L., Cabot A. *et al.* (2015). Prospects of nanoscience with nanocrystals, *ACS Nano*, 9, 1012–1057.

184. Talapin D. V., Lee J.-S., Kovalenko M. V. and Shevchenko E. V. (2010). Prospects of colloidal nanocrystals for electronic and optoelectronic applications, *Chem. Rev.*, 110, 389–458.

185. Kershaw S. V., Susha A. S. and Rogach A. L. (2013). Narrow bandgap colloidal metal chalcogenide quantum dots: Synthetic methods, heterostructures, assemblies, electronic and infrared optical properties, *Chem. Soc. Rev.*, 42, 3033–3087.

186. Kagan C. R. and Murray C. B. (2015). Charge transport in strongly coupled quantum dot solids, *Nat. Nanotechnol.*, 10, 1013–1026.

187. Shabaev A., Efros A. L. and Efros A. L. (2013). Dark and photo-conductivity in ordered array of nanocrystals, *Nano Lett.*, 13, 5454–5461.

188. Fafarman A. T., Koh W.-K., Diroll B. T. *et al.* (2011). Thiocyanate-capped nanocrystal colloids: Vibrational reporter of surface chemistry and solution-based route to enhanced coupling in nanocrystal solids, *J. Am. Chem. Soc.*, 133, 15753–15761.

189. Kovalenko M. V., Scheele M. and Talapin D. V. (2009). Colloidal nanocrystals with molecular metal chalcogenide surface ligands, *Science*, 324, 1417–1420.

190. Nag A., Kovalenko M. V., Lee J.-S. *et al.* (2011). Metal-free inorganic ligands for colloidal nanocrystals: S2-, hs-, se2-, hse-, te2-, hte-, tes32-, oh-, and nh2- as surface ligands, *J. Am. Chem. Soc.*, 133, 10612–10620.

191. Dolzhnikov D. S., Zhang H., Jang J. *et al.* (2015). Composition-matched molecular "solders" for semiconductors, *Science*, 347, 425–428.

192. Baumgardner W. J., Whitham K. and Hanrath T. (2013). Confined-but-connected quantum solids via controlled ligand displacement, *Nano Lett.*, 13, 3225–3231.

193. Sandeep C. S. S., Azpiroz J. M., Evers W. H. *et al.* (2014). Epitaxially connected pbse quantum-dot films: Controlled neck formation and optoelectronic properties, *ACS Nano*, 8, 11499–11511.

194. Evers W. H., Schins J. M., Aerts M. *et al.* (2015). High charge mobility in two-dimensional percolative networks of pbse quantum dots connected by atomic bonds, *Nat. Commun.*, 6.

195. Whitham K., Yang J., Savitzky B. H. *et al.* (2016). Charge transport and localization in atomically coherent quantum dot solids, *Nat. Mater.*, 15, 557–+.

196. Hu J. T., Li L. S., Yang W. D. *et al.* (2001). Linearly polarized emission from colloidal semiconductor quantum rods, *Science*, 292, 2060–2063.

197. Amit Y., Faust A., Lieberman I., Yedidya L. and Banin U. (2012). Semiconductor nanorod layers aligned through mechanical rubbing, *Phys. Status Solidi A*, 209, 235–242.

198. Aubert T., Paangetic L., Mohammadimasoudi M. *et al.* (2015). Large-scale and electroswitchable polarized emission from semiconductor nanorods aligned in polymeric nanofibers, *ACS Photonics*, 2, 583–588.

199. Liu Q., Cui Y., Gardner D. *et al.* (2010). Self-alignment of plasmonic gold nanorods in reconfigurable anisotropic fluids for tunable bulk metamaterial applications, *Nano Lett.*, 10, 1347–1353.

200. Perez-Juste J., Rodriguez-Gonzalez B., Mulvaney P. and Liz-Marzan L. M. (2005). Optical control and patterning of gold-nanorod-poly(vinyl alcohol) nanocomposite films, *Adv. Funct. Mater.*, 15, 1065–1071.

201. Klinkova A., Choueiri R. M. and Kumacheva E. (2014). Self-assembled plasmonic nanostructures, *Chem. Soc. Rev.*, 43, 3976–3991.

202. Tao A., Sinsermsuksakul P. and Yang P. (2007). Tunable plasmonic lattices of silver nanocrystals, *Nat. Nanotechnol.*, 2, 435–440.

203. Fan Z. and Govorov A. O. (2010). Plasmonic circular dichroism of chiral metal nanoparticle assemblies, *Nano Lett.*, 10, 2580–2587.

204. Querejeta-Fernandez A., Chauve G., Methot M., Bouchard J. and Kumacheva E. (2014). Chiral plasmonic films formed by gold nanorods and cellulose nanocrystals, *J. Am. Chem. Soc.*, 136, 4788–4793.

205. Querejeta-Fernandez A., Kopera B., Prado K. S. *et al.* (2015). Circular dichroism of chiral nematic films of cellulose nanocrystals loaded with plasmonic nanoparticles, *ACS Nano*, 9, 10377–10385.

206. Lu F., Tian Y., Liu M. *et al.* (2013). Discrete nanocubes as plasmonic reporters of molecular chirality, *Nano Lett.*, 13, 3145–3151.

207. Urban A. S., Shen X., Wang Y. *et al.* (2013). Three-dimensional plasmonic nanoclusters, *Nano Lett.*, 13, 4399–4403.

208. Alvarez-Puebla R. A., Agarwal A., Manna P. *et al.* (2011). Gold nanorods 3d-supercrystals as surface enhanced raman scattering spectroscopy substrates for the rapid detection of scrambled prions, *Proc. Natl. Acad. Sci. U. S. A.*, 108, 8157–8161.

209. Ye X., Chen J., Diroll B. T. and Murray C. B. (2013). Tunable plasmonic coupling in self-assembled binary nanocrystal superlattices studied by correlated optical microspectrophotometry and electron microscopy, *Nano Lett.*, 13, 1291–1297.

210. Liu Y., Zhou J., Zhou L. *et al.* (2016). Self-assembled structures of polyhedral gold nanocrystals: Shape-directive arrangement and structure-dependent plasmonic enhanced characteristics, *RSC Advances*, 6, S7320–S7326.

211. Nie S. M. and Emery S. R. (1997). Probing single molecules and single nanoparticles by surface-enhanced raman scattering, Science, 275, 1102–1106.

212. Zhu Z., Meng H., Liu W. *et al.* (2011). Superstructures and sers properties of gold nanocrystals with different shapes, *Angew. Chem. Int. Ed. Engl.*, 50, 1593–1596.

213. Hsu S.-W., Ngo C. and Tao A. R. (2014). Tunable and directional plasmonic coupling within semiconductor nanodisk assemblies, Nano Lett., 14, 2372–2380.

214. Young K. L., Ross M. B., Blaber M. G. *et al.* (2014). Using DNA to design plasmonic metamaterials with tunable optical properties, *Adv. Mater.*, 26, 653–659.

215. Park D. J., Zhang C., Ku J. C. *et al.* (2015). Plasmonic photonic crystals realized through DNA-programmable assembly, *Proc. Natl. Acad. Sci. U. S. A.*, 112, 977–981.

216. Ross M. B., Ku J. C., Vaccarezza V. M., Schatz G. C. and Mirkin C. A. (2015). Nanoscale form dictates mesoscale function in plasmonic DNA-nanoparticle superlattices, *Nat. Nanotechnol.*, 10, 453–458.

217. Lin Q.-Y., Li Z., Brown K. A. *et al.* (2015). Strong coupling between plasmonic gap modes and photonic lattice modes in DNA-assembled gold nanocube arrays, *Nano Lett.*, 15, 4699–4703.

218. Pompa P. P., Martiradonna L., Della Torre A. *et al.* (2006). Metal-enhanced fluorescence of colloidal nanocrystals with nanoscale control, *Nat. Nanotechnol.*, 1, 126–130.

219. Chen J., Dong A., Cai J. *et al.* (2010). Collective dipolar interactions in self-assembled magnetic binary nanocrystal superlattice membranes, *Nano Lett.*, 10, 5103–5108.

220. Chen J., Ye X., Oh S. J. *et al.* (2013). Bistable magnetoresistance switching in exchange-coupled cofe2o4-fe3o4 binary nanocrystal superlattices by self-assembly and thermal annealing, *ACS Nano*, 7, 1478–1486.

221. Zeng H., Li J., Liu J. P., Wang Z. L. and Sun S. H. (2002). Exchange-coupled nanocomposite magnets by nanoparticle self-assembly, *Nature*, 420, 395–398.

222. Urban J. J., Talapin D. V., Shevchenko E. V., Kagan C. R. and Murray C. B. (2007). Synergismin binary nanocrystal superlattices leads to enhanced p-type conductivity in self-assembled pbte/ag-2 te thin films, *Nat. Mater.*, 6, 115–121.

223. Dong A., Chen J., Vora P. M., Kikkawa J. M. and Murray C. B. (2010). Binary nanocrystal superlattice membranes self-assembled at the liquid-air interface, *Nature*, 466, 474–477.

224. Cargnello M., Johnston-Peck A. C., Diroll B. T. *et al.* (2015). Substitutional doping in nanocrystal superlattices, *Nature*, 524, 450–+.

225. Kleijn S. E. F., Lai S. C. S., Koper M. T. M. and Unwin, P. R. (2014). Electrochemistry of nanoparticles, *Angew. Chem. Int. Ed. Engl.*, 53, 3558–3586.

226. Wang C., Yin H., Dai S. and Sun S. (2010). A general approach to noble metal-metal oxide dumbbell nanoparticles and their catalytic application for co oxidation, *Chem. Mat.*, 22, 3277–3282.

227. Kang Y., Ye X., Chen J. *et al.* (2013). Design of pt-pd binary superlattices exploiting shape effects and synergistic effects for oxygen reduction reactions, *J. Am. Chem. Soc.*, 135, 42–45.

228. Kang Y., Ye X., Chen J. *et al.* (2013). Engineering catalytic contacts and thermal stability: Gold/iron oxide binary nanocrystal superlattices for co oxidation, *J. Am. Chem. Soc.*, 135, 1499–1505.

229. Yamada Y., Tsung C.-K., Huang W. *et al.* (2011). Nanocrystal bilayer for tandem catalysis, *Nat. Chem.*, 3, 372–376.

230. Zhang S., Shao Y., Yin G. and Lin Y. (2010). Electrostatic self-assembly of a pt-around-au nanocomposite with high activity towards formic acid oxidation, *Angew. Chem. Int. Ed. Engl.*, 49, 2211–2214.

231. Kang Y., Li M., Cai Y. *et al.* (2013). Heterogeneous catalysts need not be so "heterogeneous": Monodisperse pt nanocrystals by combining shape-controlled synthesis and purification by colloidal recrystallization, *J. Am. Chem. Soc.*, 135, 2741–2747.

232. Chen C., Kang Y., Huo Z. *et al.* (2014). Highly crystalline multimetallic nanoframes with three-dimensional electrocatalytic surfaces, *Science*, 343, 1339–1343.

233. Courty A., Mermet A., Albouy P. A., Duval E. and Pileni M. P. (2005). Vibrational coherence of self-organized silver nanocrystals in f.C.C. Supra-crystals, *Nat. Mater.*, 4, 395–398.

234. Poyser C. L., Czerniuk T., Akimov A. *et al.* (2016). Coherent acoustic phonons in colloidal semiconductor nanocrystal superlattices, *ACS Nano*, 10, 1163–1169.

235. Podsiadlo P., Krylova G., Lee B. *et al.* (2010). The role of order, nanocrystal size, and capping ligands in the collective mechanical response of three-dimensional nanocrystal solids, *J. Am. Chem. Soc.*, 132, 8953–8960.

236. Schmitt M., Choi J., Hui C. M. *et al.* (2016). Processing fragile matter: Effect of polymer graft modification on the mechanical properties and processibility of (nano-). particulate solids, *Soft Matter*, 12, 3527–3537.

237. Gauvin M., Wan Y., Arfaoui I. and Pileni M.-P. (2014). Mechanical properties of au supracrystals tuned by flexible ligand interactions, *J. Phys. Chem. C*, 118, 5005–5012.

238. Salerno K. M., Bolintineanu D. S., Lane J. M. D. and Grest G. S. (2015). Ligand structure and mechanical properties of single-nanoparticle-thick membranes, *Physical Review E*, 91.
239. Klajn R., Bishop K. J. M., Fialkowski M. *et al.* (2007). Plastic and moldable metals by self-assembly of sticky nanoparticle aggregates, *Science*, 316, 261–264.
240. Bigioni T. P., Lin X. M., Nguyen T. T. *et al.* (2006). Kinetically driven self assembly of highly ordered nanoparticle monolayers, *Nat. Mater.*, 5, 265–270.
241. Mueggenburg K. E., Lin X. M., Goldsmith R. H. and Jaeger H. M. (2007). Elastic membranes of close-packed nanoparticle arrays, *Nat. Mater.*, 6, 656–660.
242. Kanjanaboos P., Lin X.-M., Sader J. E. *et al.* (2013). Self-assembled nanoparticle drumhead resonators, *Nano Lett.*, 13, 2158–2162.
243. Wang Y., Liao J., Mcbride S. P. *et al.* (2015). Strong resistance to bending observed for nanoparticle membranes, *Nano Lett.*, 15, 6732–6737.
244. Wang Y., Kanjanaboos P., Barry E. *et al.* (2014). Fracture and failure of nanoparticle monolayers and multilayers, *Nano Lett.*, 14, 826–830.
245. Van Rijn P., Tutus M., Kathrein C. *et al.* (2013). Challenges and advances in the field of self-assembled membranes, *Chem. Soc. Rev.*, 42, 6578–6592.
246. He J., Lin X.-M., Chan H. *et al.* (2011). Diffusion and filtration properties of self-assembled gold nanocrystal membranes, *Nano Lett.*, 11, 2430–2435.
247. Barry E., Mcbride S. P., Jaeger H. M. and Lin X.-M. (2014). Ion transport controlled by nanoparticle-functionalized membranes, *Nat. Commun.*, 5.
248. Jang B., Park M., Chae O. B. *et al.* (2012). Direct synthesis of self-assembled ferrite/carbon hybrid nanosheets for high performance lithium-ion battery anodes, *J. Am. Chem. Soc.*, 134, 15010–15015.

Chapter 2

Visible-Light Nanocatalyst Based on Porphyrin Self-Assembly

Li Qi and Bai Feng

2.1 Introduction

With the increasing attention on environmental issues and clean energy from water and sunlight, photochemical conversion of solar energy has been found broad potential applications, especially biomimetic materials with the ability to promote visible-light harvesting efficiency and photochemical reactions have been extensively studied.[1-4] Although remarkable progress has been achieved in both the synthetic strategy and preliminary photocatalytic mechanism for advanced biomimetic material, it is still a big challenge to realize practical applicable efficiency for visible-light photocatalytic reaction, even directly utilizing natural sunlight. Much more efforts are still required to understand the basic interaction processes of photons and electrons in the related biomimetic photocatalytic materials, and to the basic laws for the relationships between these biomimetic materials and their photocatalytic behaviors.

Porphyrin, attractive building blocks for biomimetic photocatalyst, has similar molecular structure to photoactive molecules chlorophyll, and provides analogous chemical properties to the natural photocatalyst for biological energy transduction processes in plants, algae.[5,6] Porphyrin is also one of the most important visible-light harvesting materials because of its rich and extensive visible-light absorption features. Light harvesting is the first step for the whole photocatalytic process, and the number and energy of photons absorbed by the photocatalysts restrict their photocatalytic nature.[7,8] Especially, porphyrin organic material has a long lifespan in the excited state as well as efficient photoabsorption in the visible-light region, and the long lifespan of porphyrin in the excited state allows for effective charge separation. So porphyrin is one of the most promising organic visible-light antenna candidates for optical and electronic applications.[9]

In biological systems, tetrapyrroles such as porphyrins and chlorophylls are often self-organized into nanoscale superstructures that perform many of the essential light-harvesting and energy- and electron-transfer functions.[10] Porphyrin self-assembly is structurally similar to natural porphyrin-type light-harvesting and photosynthesis pigments, and might be expected to mimic important aspects of natural photosynthetic systems.[11,12] Self-assembly process is a natural and spontaneous process occurring mainly through non-covalent bonds and interactions such as delocalized conjugated π structures, van der Waals force, hydrogen bond, metal-coordination bond and electrostatic interaction.[13–15] Due to their rapid photoinduced charge separation and a relatively slow charge recombination, delocalized conjugated π structure material has been extensively researched in electron-transfer processes and act as a novel structure, which could form a common conjugated system and improve the photocatalytic activity.[16,17] The porphyrin self-assembly, which assembles the modular units into specific spatial arrangements and facilitates them to work cooperatively, is vital for the achievement of photofunctions in photovoltaic systems.[18] Thus, the porphyrin self-assembly with conjugated π structure is expected to exhibit more complicated absorption and emission spectra than those of the monomers.[19–21] A rich variety of nanoscale assemblies of porphyrins including flower, rods, tubes and particles have been organized by aforesaid non-covalent bonds and interactions.[22–27] These porphyrin assemblies with well-defined structures can photoinduce electron transfer, charge transport and photochemical reactions.

Until now, many research papers have been published on porphyrin self-assembly as a visible-light catalyst, this section aims at summarizing the recent research breakthroughs in morphology control and methodologies for fabrication of porphyrin bionic self-assembly, as well as their application performance in visible-light photocatalytic water splitting and visible-light photodegradation of organic pollutants. Great emphasis will be attached to the emerging strategies for the visible-light activity improvement of porphyrin assembly and better understanding of the fundamental mechanisms. Finally, we are trying to provide guidelines for the rational design and fabrication of highly efficient porphyrin self-assembly based nanomaterials for visible-light photocatalysis.

2.2 Methodologies for Fabrication of Porphyrin Self-assembly

Self-assembly is a ubiquitous principle in nature, which can lead to ordered architectures, and can occur in natural and synthetic systems at various levels.[28] It is currently considered to be an efficient "bottom-up" strategy for manufacturing porphyrin-based self-assembly nanomaterials through non-covalent bonds and interactions such as van der Waals force, hydrogen bond, electrostatic interaction and metal-coordination bonding. Various self-assembly protocols, including surfactant-assisted self-assembly (SAS), microemulsion self-assembly (MS), micelle confined self-assembly (MCS), metal coordination-assisted self-assembly (MCAS), have been utilized to formulate numerous porphyrin nanostructure assemblies. Porphyrin nanostructures with a certain size, shape and function can be provided through careful molecular or supramolecular design and precise control of various intermolecular noncovalent interactions.

2.2.1 *Surfactant-assisted Self-assembly*

Among various methodologies of formulate porphyrin self-assembly, SAS is most attractive strategy, because the nature and morphologies of the assembled nanomaterials can readily be regulated by surfactants. In the SAS process, porphyrin molecules dissolved in a guest solvent are organized with the assistance of surfactants that are dispersed in a host solvent.[28,29] Accordingly to intersolubility of the guest and host solvent, two types of SAS methods are developed to construct porphyrin assembly.

2.2.1.1 *Incompatible Solvent SAS Method*

Organic porphyrin solution is added drop wise into an aqueous surfactant solution. Hollow nanospheres, solid nanospheres, nanotubes, nanorods, and nanofibers of porphyrin have been successfully prepared by carefully controlling the assembly behavior with SAS method. The porphyrin nanoassemblies can show different chiroptical, photocatalytic performance and exhibit morphology-dependent photocatalytic efficiency.[28,30,31] For example, porphyrin-based hollow hexagonal nanoprisms have been successfully prepared by Wan and co-workers in a dimethyl formamide/water (DMF/water) system in the presence of cetyltrimethylammonium bromide (CTAB) surfactant.[31] These hollow nanostructures could be further organized into an ordered three-dimensional (3D) architecture. Both sodium dodecyl benzene sulfonate (SDBS) and myristyltrimethylammonium bromide (MTAB) aqueous are also employed in the SAS method for formation of various porphyrin nanostructures.[30,35]

2.2.1.2 *Compatible Solvent SAS Method*

It is most common for SAS method that guest and host solvents have similar polarity and good compatibility. In this case, porphyrin nanostructure assemblies are generally produced in a mixed solvent system where the employed solvents have good compatibility.[36,21] Hasobe and co-workers reported that Fullerene has also been encapsulated porphyrin hexagonal nanorods in a DMF/acetonitrile system mixed with surfactant.[13,37]

These above investigations suggest that the SAS method using an oil/aqueous medium or compatible medium is an efficient and controlled way to construct porphyrin assembly-based nanomaterials which can show different assembly properties.

2.2.2 *Microemulsion Self-assembly*

Interfacial self-assembly driven microemulsion is a facile and efficient method for rationally designing and fabricating molecular assemblies with directing structure and property control over multiple scales.[38–40] MS has been extended to fabricate porphyrin nanocrystals from optically active precursor porphyrin by our group.[41,35] MS is fabricated dispersion of an organic solution containing porphyrin molecules to another immiscible aqueous solution. And subsequently, the organic solvent is removed through thermally driven evaporation, which induces the droplets shrinking and the porphyrin concentration therein rises. And then, porphyrin molecules self-assemble into nanostructure within mircoemulsion droplets.[42] As shown in Figure 2.1(a), porphyrin moleculars are self-assembled into various hierarchically ordered nanocrystals through an interfacial self-assembly driven microemulsion process.[35] The resulted nanocrystals display uniform shapes and sizes tuned from tens to few hundred nanometers in Figures 2.1(b)–2.1(d).[35]

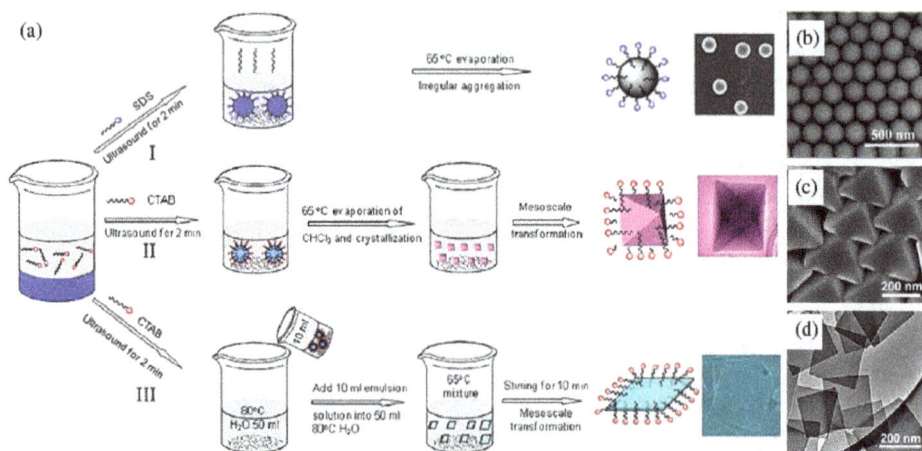

Figure 2.1   (a) Schematic illustration of the synthesis processes of porphyrin nanocrystals by MS. Structure and morphology of the hierarchically structured porphyrin nanocrystals: (b) scanning electron microscope (SEM) image of the self-assembled microspheres, (c) representative SEM image of the self-assembled octahedral arrays, (d) TEM image of the self-assembled nanosheets. Adapted with permission from Ref. 35, copyright 2014 American Chemical Society.

2.2.3 *Micelle Confined Self-assembly*

Micelle confined self-assembly has been adopted to form self-assembled nanocrystal superlattices with alkanethiol-stabilized gold nanocrystals as a building block in our previous work.[43,44] Bai and his colleagues have extended this self-assembly induced micelle encapsulation synthetic method to fabricate porphyrin nanocrystals using the optically active precursor zinc-tetra(4-pyridyl) porphyrin (ZnTPP).[29] Using this method, metal porphyrins are self-assembled into highly ordered 3D arrays through combined π–π interaction and coordination of the peripheral pyridine groups to the core metal ions. Through confined non-covalent interactions within surfactant micelles, zinc meso-tetra (4-pyridyl) porphyrin (ZnTPyP) nanocrystals with a series of morphologies including nanodisks, tetragonal nanorods, hexagonal nanorods, as well as amorphous spherical particles are synthesized with controlled size and dimension. The morphologies and size distribution are shown in Figure 2.2.[29]

Bai's group has summarized a phase diagram of porphyrin self-assembly nanostructures in Figure 2.3, illustrating that morphology control is achieved *via* kinetically controlled nucleation and growth with MCS.[29] From formation morphologies and size distribution of aforementioned ZnTPyP assemblies in Figure 2.2, solution pH and surfactant concentration are two key factors influencing the nanocrystal nucleation and growth in reactions with a given concentration of porphyrin. Within each individual phase region, along with increasing of surfactant concentration, the nanocrystal dimension grows. They have observed that the nanocrystal dimension also increases when pH value rises. At given pH conditions, along with growth of surfactant concentration, one can achieve morphology from irregular particles to those that have well-defined morphologies including disks, tetragonal rods, and hexagonal rods. In other words, these nanocrystals with controlled dimension and morphology can be prepared by MCS.

2.2.4 *Metal Coordination-assisted Self-assembly*

Coordination bond is one of the controllable factors for design and preparation of porphyrin self-assembly with metal ions and organic ligands. Cooperation and competition between additionally introduced coordination interaction and the original intermolecular interaction can lead to self-assembly of functional molecular materials into novel nanostructures with different morphology. As shown in Figure 2.4, porphyrin assemblies such as nanocubes, nanorods and microrods have been successfully prepared by metal coordination-assisted self-assembly.[45,46] According to X-Ray diffraction (XRD) patterns and infrared absorption spectra, a schematic illustration of the proposed structures of 5,10,15,20-tetrakis-(4-carboxyphynyl)-21H,23H porphyrin (H2TCPP) assemblies is summarized in

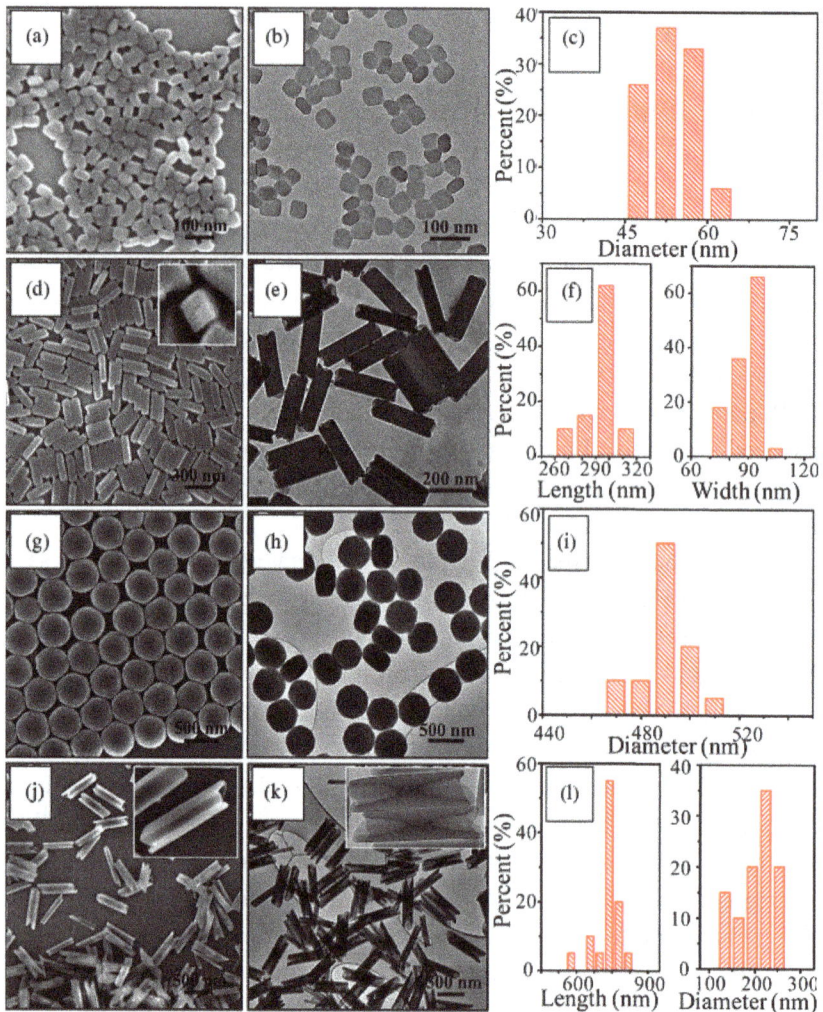

Figure 2.2 Representative SEM images (first column), TEM images (second column), and size distribution (third column) of porphyrin nanocrystals with different morphologies nanoparticles (a)–(c), tetragonal nanorods (d)–(f), hexagonal porous nanodisks (g)–(i), and hexagonal nanorods (j)–(l). Reprinted with permission from Ref. 29, copyright 2014 American Chemical Society.

Figure 2.4(e). The single crystal and molecular structure of MCAS porphyrin self-assemblies revealed by XRD analysis for both porphyrin derivatives renders it possible to investigate the formation mechanism as well as the molecular packing conformation of self-assembled nanostructures of these typical organic building blocks with large conjugated system in a more confirmed manner.

MCAS has been expanded to porphyrin metal-organic framework (Porphyrin-MOF) thin films. MOFs have attracted considerable attention and been applied in

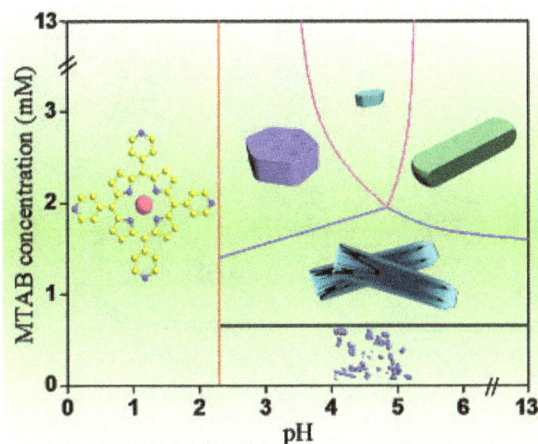

Figure 2.3 Nanocrystal morphology evolution with the surfactant concentration and pH. Reprinted with permission from Ref. 29, copyright 2014 American Chemical Society.

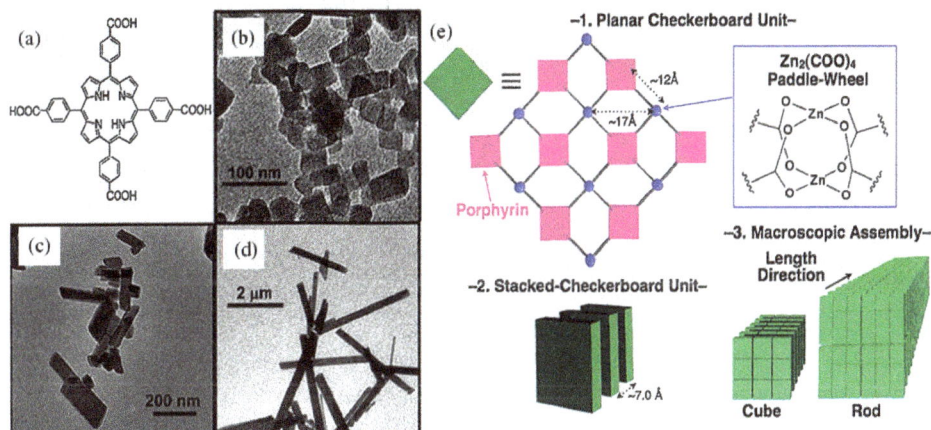

Figure 2.4 (a) Chemical structure of H_2TCPP and transmission electron microscopy (TEM) images of (b) nanocubes (52.5 mM), (c) nanorods (105 mM) and (d) microrods (420 mM). (e) A schematic illustration of the proposed structures of H_2TCPP architectures *via* paddle-wheel complexes in this study. Adapted with permission from Ref. 46, copyright 2012 Royal Society of Chemistry.

catalysis due to high surface areas and adjustable holes. Porphyrins show high activities in homogeneous catalysis, but they are unstable and prone to degradation in the catalytic oxidation process, which limit their practical applications. Porphyrin-MOF exhibit special catalytic activity sustainably by avoiding degradation. MnPorphyrin-MOF thin film consisting of [5,10,15,20-tetra(4-carboxyphenyl)porphyrin]Mn(III) (MnTCPP), zinc(II) acetate (Zn^{2+}) and

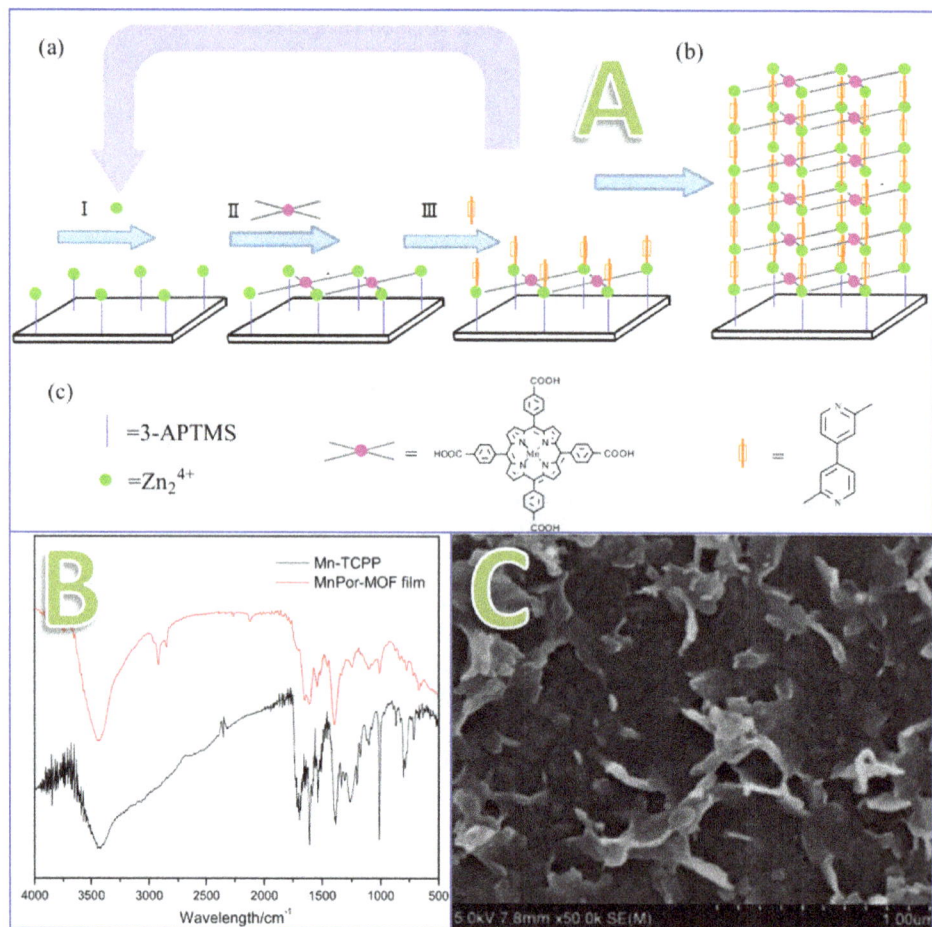

Figure 2.5 (A) (a, b) Step-by-step growth of MnPor-MOF, (c) representations of building blocks used in fabrication of MnPor-MOF film. (B) IR spectra of MnTCPP and MnPor-MOF film. (C) FE-SEM images of MnPor-MOF grown on 3-APTMS-functionalized quartz glass. Adapted with permission from Ref. 47, copyright, 2015 John Wiley & Sons, Ltd.

2,2′-imethyl-4,4′-bipyridine (dmbpy) has been prepared on a functionalized quartz glass surface by the layer-by-layer self-assembly technique drive by metal coordination-assisted self-assembly, and the fabrication process shows in Figure 2.5.[47] The field emission scanning electron microscopy (FE-SEM) image (Figure2.5(c)) shows that form dense and smooth of Mnporphyrin-MOF film. Infrared absorption (IR) spectra of MnPor-MOF film shows a shift of the carboxylate stretching band to lower frequency owing to the transformation of carboxylic acid groups in porphyrin to carboxylate chelating functionality. These results has revealed porphyrin-MOF film can be successfully fabricated by metal coordination-assisted self-assembly.

Intermolecular metal-ligand axial coordination is another approach for metal coordination-assisted self-assembly. A vast majority of the metals are capable of forming metal porphyrins with tetrapyrroles of square planar geometry. And many central metal atoms of these metal porphyrin are unsaturated thus exhibiting their desire for additional one or two ligands. This metal-ligand axial interaction provides moderate to high stability for the resulting supramolecular light energy harvesting systems.[48]

2.3 Photocatalysis Properties and Applications

Light harvesting is the first and critical step for the whole photocatalytic process, and the number and energy of photons absorbed by the photocatalyst restrict its own photocatalytic activity.[49] Because of the strong absorption in the region of 400–450 nm (one Soret band), as well as 500–700 nm (four Q bands), porphyrin derivatives have already been successfully used for photocatalysis, and used as main component of light-harvesting systems in both photosynthetic bacteria and green plants.[24,7] Porphyrin molecules as organic photocatalysts are limited by photostability, hence incurs a gradual loss in their photocatalytic efficiency during the reaction. The geometrical constraints imposed by the rigid aggregate framework will therefore make attack by reactive species more difficult and ensure enhanced stability for further use.[50] Moreover, owing to porphyrin molecular unique planar, rigid molecular geometry, and aromatic electron delocalization over the molecular frame, porphyrin assemblies are endowed visible-light photostability as well as sustainable photocatalytic activity, such as visible-light photodegradation of organic pollutants, visible-light photocatalytic water splitting.

2.3.1 *Visible-light Photodegradation of Organic Pollutants*

One meaningful reaction of porphyrin assembly-based nanomaterials is photocatalytic degradation organic pollutants. Most of the organic pollutants, such as methyl orange (MO), rhodamine B (Rh B) and methylene blue (MB), can be photodegraded by porphyrin assembly-based photocatalyst under visible-light irradiation. As exhibited in Figure 2.6, similar to the semiconductor material for photocatalytic degradation of pollutants, porphyrin assembly absorbs visible-light to induce π–π* transition and transport the excited-state electrons from the highest occupied molecular orbital (HOMO) to the lowest unoccupied molecular orbital (LUMO). Two kinds of oxidize mechanisms has been proposed to explain the photocatalytic degradation activity. The first photocatalytic degradation pollutants mechanism is that photogenerated holes of the HOMO can directly oxidize the organic pollutants to form oxidation products. The second mechanism is

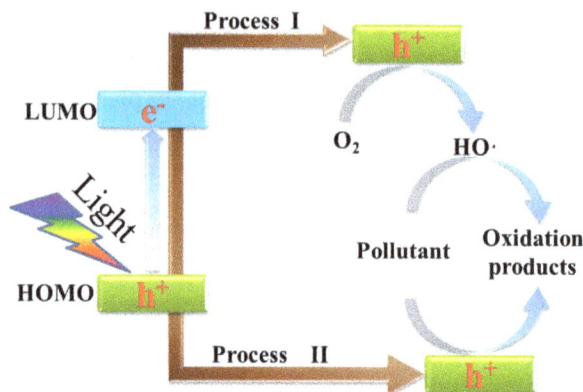

Figure 2.6   Two processes of photocatalytic degradation pollutants: (Process I), holes of HOMO can directly oxidize the organic pollutants to form oxidation products. (Process II), photoinduction electron can also react with surface adsorbed H_2O and/or O_2 to produce hydroxyl radicals, which are strong oxidizing agents and can oxidize almost all organic pollutants with no selectivity.

that photoinduction electron react with surface adsorbed O_2 to produce hydroxyl radicals, which are strong oxidizing agents and can oxidize almost all kinds of organic pollutants with no selectivity.

2.3.1.1 *Direct Oxidizing Mechanism for Visible-light Photodegradation Organic Pollutants*

Photocatalytic activities of different porphyrin assemblies for photodegradation pollutant under the visible-light irradiation have been demonstrated and photodegradation mechanism is shown in Figure 2.6. And direct oxidizing mechanism has been given to explain the photocatalytic degradation activity in Figure 2.7. Different morphologies of meso-tetra(4-carboxyphenyl) porphyrin (TCPP) assemblies (spherical to flower shaped, J-aggregated assemblies) have been synthesized using SAS methed.[51] There are strong intermolecular π–π interaction between aggregated porphyrin molecules and aggregated porphyrin assemblies with delocalized conjugated π structures have been extensively studied in electron-transfer processes due to their rapid photoinduced charge separation and a relatively slow charge recombination, which are good photo semiconductors.[17] As shown in Figures 2.6 and 2.7, Rh B, which is a common water pollutant demonstration molecules, can be photodegraded *via* direct utilization the holes as oxidant by porphyrin self-assembly. The experiment results indicate that the photocatalytic activity depends upon the morphology of the TCPP aggregated structures and the photocatalytic efficiency of the flakes and flower shaped aggregates is lower than rod shaped TCPP aggregates.[51]

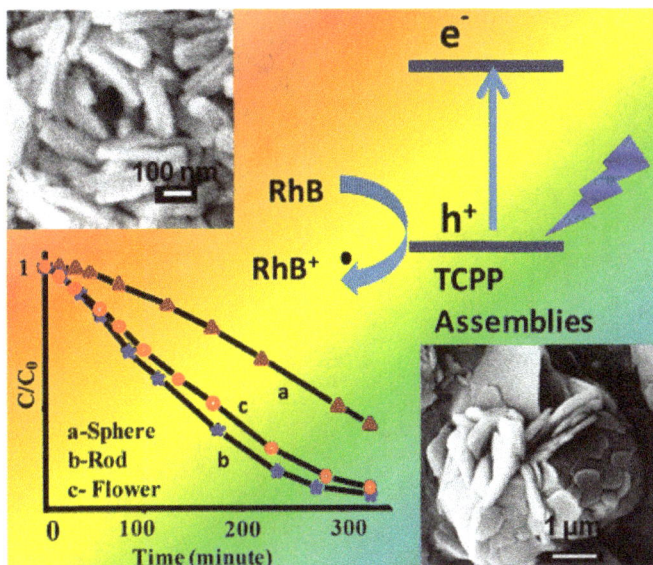

Figure 2.7 Photocatalytic activity of surfactant-assisted different morphologies of TCPP aggregates. And photocatalytic mechanism of photodegradation RhB under the visible-light irradiation. Reprinted with permission from Ref. 51, copyright 2014 American Chemical Society.

J-aggregate porphyrin nanofiber assembly favors the electron transfer to form hydroxyl radical species and enhances photocatalytic activity. Zn-porphyrin (ZnTPyP) nanofibers are synthesized by means of SAS method shown in Figure 2.8(b) and used to photocatalytic degradation of Rh B pollutant under visible-light irradiation.[30,28] The ZnTPyP assembles display distinct photocatalytic activity for photodegradation of RhB molecules. This photocatalyst also exhibited an excellent photostability without significant decreased photocatalytic activity (Figure 2.8(c)). Hydroxyl radical species generation and electron transfer process in the nanofiber system as shown in Figures 2.8(a) and 2.6 are both responsible for the superior photocatalytic performance of the J-aggregates ZnTPyP fibrous nanoassemblies. The photocatalytic activity of the ZnTPyP nanofibers is, to a great extent, not related to an energy transfer mechanism but to an electron transfer process.[52,30,17] Strong intermolecular π electronic coupling between the ZnTPyP chromophores exist, which are cooperatively aligned as J-aggregates with a slipped cofacial arrangement.[53,7] This could favor the coherent electronic delocalization over the aggregated chromophores, making these ZnTPyP-based nanofibers efficient photosemiconductors, wherein the photoinduced electron transfer could be facilitated *via* the coherently delocalized π electrons.[30] Accordingly, it is acceptable that the photocatalytic performance of ZnTPyP nanofibers mainly relies on an electron transfer process.

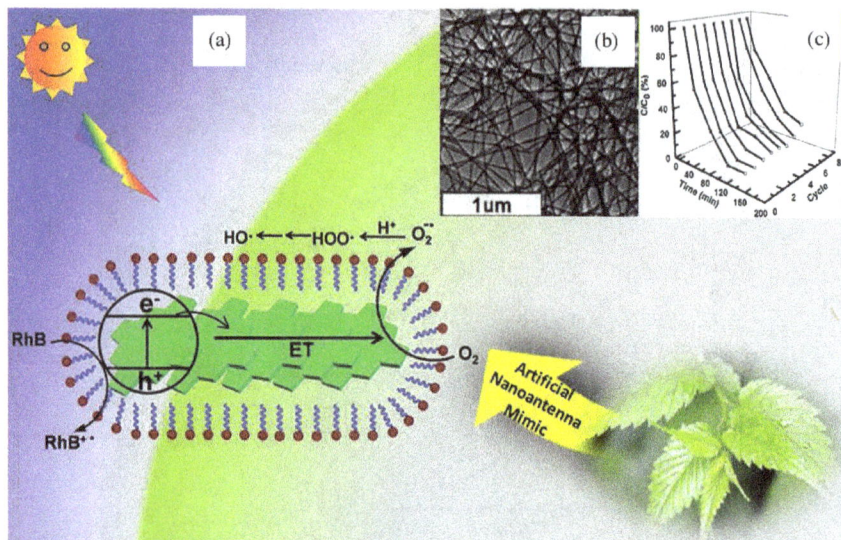

Figure 2.8   (a) Schematic illustration on the structure of ZnTPyP nanofibers, and their morphology-dependent photocatalytic performance. ET = electron transfer. (b) Image of the ZnTPyP nanofibers. (c) Eight consecutive cycling photodegradation curves of RhB over ZnTPyP nanofibers under visible-light irradiation. Adapted with permission from Ref. 30, copyright 2012 Royal Society of Chemistry.

2.3.1.2 *Radicals Oxidizing Mechanism for Visible-light Photodegradation Organic Pollutants*

Kinetic curves of photocatalytic degradation results (Figure 2.9(a)) demonstrate that porphyrin-MOF film, H_2O_2 and visible light are essential for the degradation of methylene blue. The photocatalytic degradation reaction is studied using electron spin resonance (ESR) spectroscopy (Figure 2.9 (b)). For the system of porphyrin-MOF film, H_2O_2 and DMPO, a hydroxyl radical trapping agent, a signal of the DMPO–OH adduct is detected upon photoexcitation (Figure 2.9(b), curve A). When methylene blue is added to the system, however, the magnitude of the signal is greatly attenuated (Figure 2.9(b), curve B), indicating that •OH is consumed. Based on aforementioned results obtained, a plausible reaction process show in Figure 2.9(c) for photocatalytic degradation of methylene blue in the presence of porphyrin-MOF film. Porphyrin-MOF film accepts a photon to form an excited state under photoirradiation, and back to the ground state releases photogenerated electrons. H_2O_2 can quickly capture the photogenerated electrons to produce active species •OH, which efficiently degrades methylene blue.

2.3.2 *Visible-light Photocatalytic Water Splitting*

To tackle the energy and environmental problems, extensive research has been carried out on photocatalytic splitting of water into H_2 and O_2 since water splitting

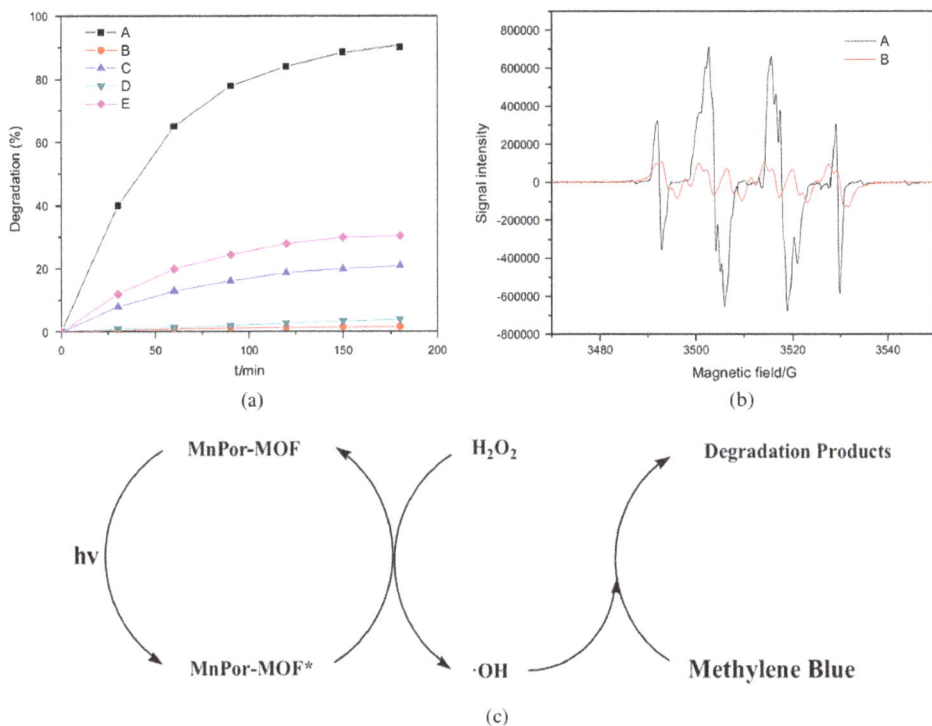

Figure 2.9 (a) Kinetic curves of photocatalytic degradation of methylene blue under various conditions at 30°C: MnPor-MOF film, H_2O_2 and visible light (curve A); MnPor-MOF film and visible light (curve B); H_2O_2 and visible light (curve C); MnPor-MOF film and H_2O_2 (curve D); MnTCPP, H_2O_2 and visible light (curve E). (b) ESR spectra of systems upon photoirradiation: MnPor-MOF film, H_2O_2 and DMPO (curve A); MnPor-MOF film, H_2O_2, methylene blue and DMPO (curve B). (c) Proposed process for the degradation of methylene blue. Adapted with permission from Ref. 47, copyright, 2015 John Wiley & Sons, Ltd.

using photoelectrochemical was first reported by Fujishima and Honda.[54] For water splitting process, water molecules are reduced by the photoinduced electrons to form H_2 and are oxidized by the photoinduced holes to produce O_2.[55] Although transition metal oxides are nontoxic materials with excellent photocatalytic properties, high chemical stability and low cost, most of transition metal oxides have a wide energy band gap and correspond to ultraviolet region. Unfortunately, natural sunlight which only contains 4% ultraviolet light would be able to effectively induce transition metal oxides photocatalysis. Therefore, a great number of efforts have been devoted to develop new photocatalysts with visible light response.[56,9]

An efficient visible-light-induced hydrogen evolution system has been developed by using supramolecular porphyrin hexagonal nanocylinders that encapsulate Pt-colloids-deposited TiO_2 (Pt/TiO_2) in the internal cavity.[57] Zinc meso-tetra(4-pyridyl) porphyrin [ZnP(Py)$_4$] assembly was formed with the SAS

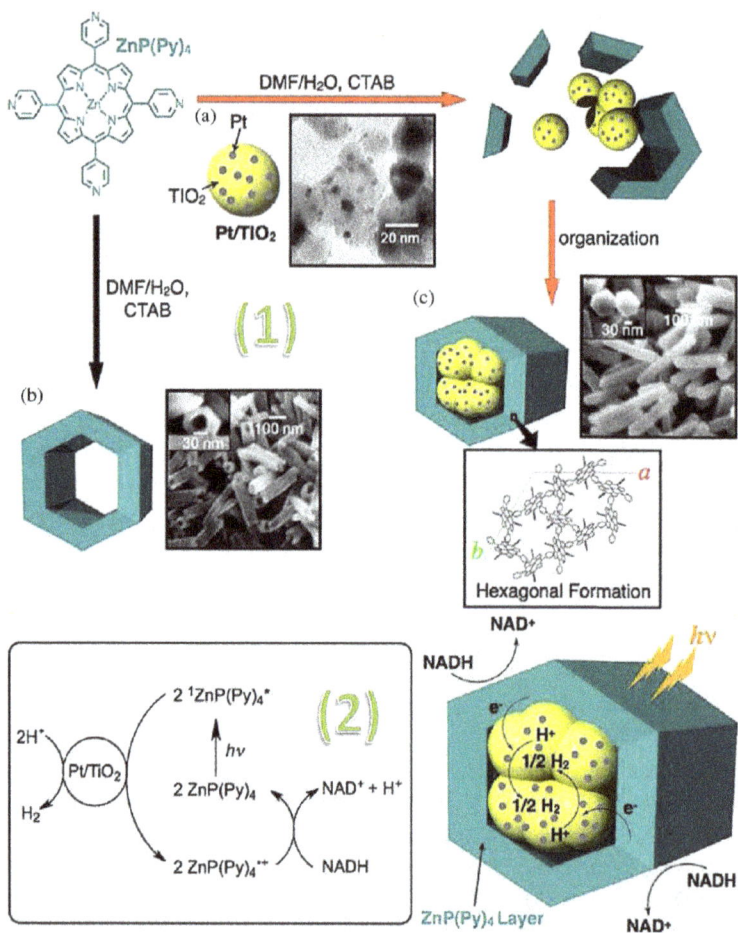

Figure 2.10   (a) A schematic illustration of organization procedure of Pt/TiO$_2$-ZnP(Py)$_4$ nanorods in this study. The electron micrograph images show (1) Pt/TiO$_2$ (TEM), (2) ZnP(Py)$_4$ nanocylinder (SEM), and (3) Pt/TiO$_2$-ZnP(Py)$_4$ nanorods (SEM). (b) Schematic Illustration for Mechanism of Hydrogen Evolution. Adapted with permission from Ref. 57, copyright 2013 American Chemical Society.

method and Pt/TiO$_2$ nanoparticles were encapsulated within a ZnP(Py)$_4$ hexagonal nanocylinder *via* a solvent mixture technique as shown in Figure 2.10(a). Pt/TiO$_2$-ZnP(Py)$_4$ also shows a broadened absorption in the visible region because of aggregation of ZnP(Py)$_4$ and exhibited efficient hydrogen evolution under visible-light irradiation, whereas no hydrogen was evolved in the case of Pt/TiO$_2$ without ZnP(Py)$_4$. The photocatalytic mechanism of the porphyrin-based composites was examined by femtosecond time-resolved transient absorption spectroscopy and schematic illustration shown in Figure 2.10(b).

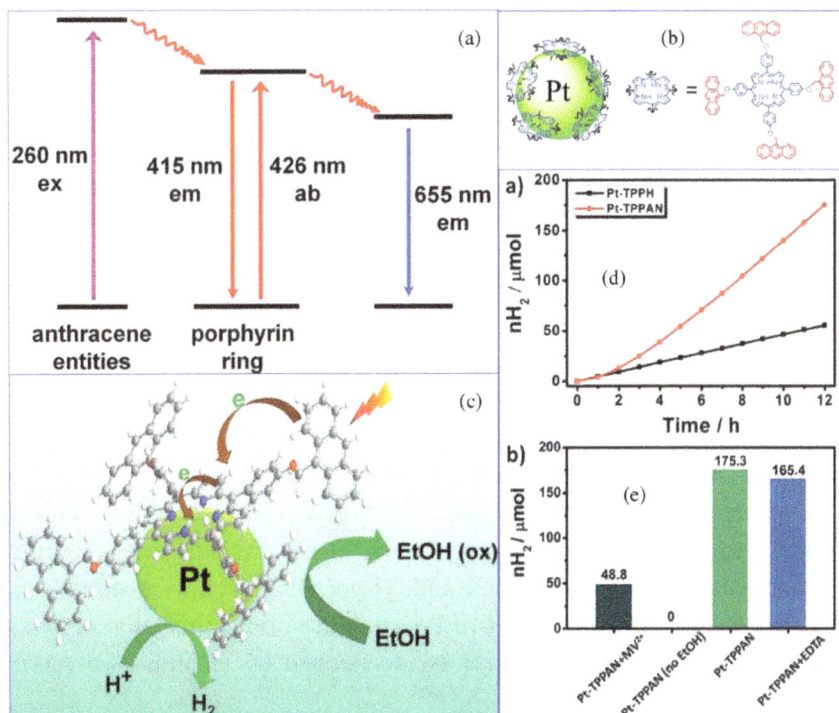

Figure 2.11 (a) Schematic diagram of the fluorescence spectrum of TPPAN excited at 260 nm at room temperature. (b) The proposed configuration of the Pt-TPPAN nanocomposite. (c) Schematic representation of the electron transfer and the photosensitized hydrogen evolution in a D-A anthryl-porphyrin functionalized Pt nanocomposite in an aqueous ethanol solution. (d) The amount of H_2 evolved as a function of irradiation time using Pt-TPPH and Pt-TPPAN nanocomposites as photo-catalysts. (e) The amount of H_2 evolved using Pt-TPPAN nanocomposites as photocatalysts with various sacrificial reductants and electron mediators under UV-Vis light irradiation for 12 hour. Adapted with permission from Ref. 58, copyright 2013 Royal Society of Chemistry).

A novel porphyrin (5, 10, 15, 20-tetrakis (4-(anthracene-1-ylmethoxy) phenyl) porphyrin, TPPAN) and its functionalized platinum nanoparticles (Pt-TPPAN) were assembled (Figure 2.11(b)). The Pt-TPPAN nanocomposite as a photocatalyst, a new and more compact photoinduced hydrogen evolution system from an aqueous ethanol solution without additives was developed.[58] Fluorescence results exhibit that energy transfers from the peripheral anthracene substituents to the porphyrin ring cause the porphyrin to fluoresce. The energy transfer process is presented in Figure 2.11(a). Photoinduced electron transfer and charge transport processes are of paramount importance, because they constitute the basis for organic photoactivity.[18,58] It may be described as in Figure 2.11(c) that a possible pathway of the photoinduced hydrogen production from this new and simple system using an anthryl-porphyrin functionalized platinum nanocomposite as the

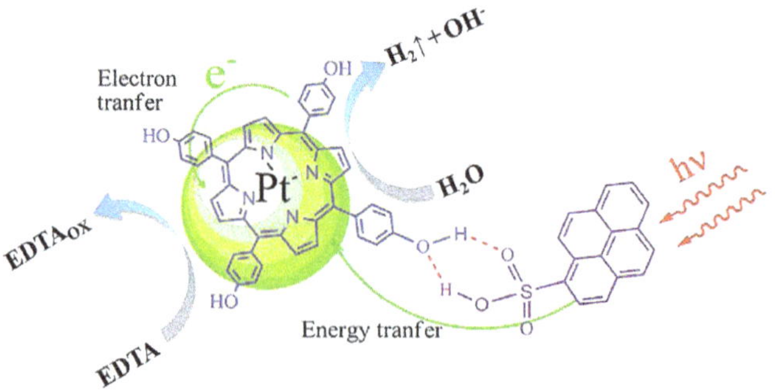

Figure 2.12   Proposed mode of energy and electron transfer from porphyrin-pyrene into a Pt nanoparticle. Adapted with permission from Ref. 60, copyright 2011 Science Direct.

photocatalyst from an aqueous ethanol solution: (i) Under UV–Vis light irradiation, the energy could be transferred from the excited peripheral anthracene substituents to the porphyrin moiety to form an excited anthryl-porphyrin. (ii) Electrons from the photoexcited anthryl-porphyrin could be directly transferred to the metallic Pt nanocore. (iii) H+ is reduced on the metallic Pt nanocore forming H_2. (iv) The photosensitizer is regenerated by ethanol as a sacrificial reductant.

The photocatalytic results of the Pt-TPPAN colloids are shown in Figure 2.11(d). The occurrence of photoinduced electron transfer typically relies on coupling an electron donor with an acceptor having suitable frontier molecular orbitals. In terms of chemical structures, frequently employed donor groups are electron-rich conjugated structures. Anthracene and electron-deficient aromatic molecules are often used electron acceptors.[18,59,58] After 12 h of UV–Vis light irradiation and in the absence of an electron mediator, the total amount of H_2 evolved from the Pt-TPPAN (175.3 mmol) is much higher than that from the pristine free-based porphyrin functional Pt nanocomposite (Pt-TPPH, 55.6 mmol). Interestingly, in the presence of electron mediator, Pt-TPPAN demonstrated relative low photocatalytic activities under the same reaction conditions (Figure 2.11(e)). The anthryl-porphyrin molecules in the nanocomposite worked as light absorption antennas and the nanocore Pt species acted as co-catalyst.[58] Thus, it can be concluded that the direct electron transferring from the excited porphyrin assembly to the platinum nanoparticles can avoid energy loss or back reactions.

Similar to the above-mentioned Pt-TPPAN nanocomposite, porphyrin-pyrene conjugate functionalized Pt assembly nanocomposite has been also constructed as a photocatalyst for photoinduced hydrogen production and ethylenediaminetetraacetic acid (EDTA) as a sacrificial reductant in the absence of an electron mediator shown in Figure 2.12.[60] The photocatalytic mechanism of the porphyrin-pyrene-Pt assembly nanocomposite is in agreement with that of aforementioned porphyrin-anthracene-Pt assembly photocatalysts.[58] The photoinduced energy transfer from the photoexcited state of pyrene to the porphyrin occurs, accompanied by an electron transfer from the excited porphyrin moiety to the platinum catalyst for the H_2 generation.

Recently, the mechanism of direct photoinduced homolytic cleavage of H_2O molecules into H• and OH• radicals is theoretically proposed and confirmed by Sobolewski and Domcke for the oxotitanium porphyrin-water complex.[61,62] In this approach, the porphyrin moiety serves as the antenna for visible-light absorption, while the TiO group exhibits the redox properties needed for the oxidation of water.

2.3.3 *Visible-light Photocatalytsis in Other Applications*

2.3.3.1 *Reduction of Noble Metal*

Various porphyrin nanoassemblies can be utilized for synthesis of metals nanoparticles and networks *via* visible-light photocatalytic reduction of a metal salt precursor.[24,41,63] As shown in Figure 2.13(e), the photocatalytic reduction of noble metal salts by the porphyrin nanoassemblies has been conducted under visible-light in the presence of an electron donor (ED). And these porphyrin self-assemblies photoreactions constitute a reductive photocatalytic cycle. Through the photocatalytic reaction, noble metal salt is initially reduced to form highly uniform noble metal nanoparticle seeds in the presence of light. Then a shell of noble metal nanoparticle networks is formed at the surface of the photoactive porphyrin self-assemblies as the noble metal precursors are continuously reduced. Using four kinds of morphology porphyrin assemblies as templates, Pt nanosheets, Pt nanospheres, Pt octahedra and Pt long tubes are reduced by photocatalytic reduction and the TEM results are showed in Figures 2.13(a)–2.13(d).[35]

Co-assembly from peptide and porphyrin, simple biologically relevant molecule is also able to induce the epitaxial oriented growth of Pt nanostructure in a photocatalytic reaction. Yan and co-workers reported peptide-induced hierarchical porphyrin assemblies which consist of a network of J-aggregate nanoscale substructures that serve as light-harvesting antennae with a relatively broad spectral cross-section and considerable photostability.[24,64] The results show that photoreduction of the PtII salt occurs within 5 minutes of irradiation to produce Pt nanoparticles on the surface of the porphyrin assemblies. They pointed out that porphyrins capture light to produce photoexcited states that were rapidly reduced by an electron donor such as ascorbic acid, and radical anion can then be used catalytically to reduce various metal salts to the metallic state through a succession of light harvesting and photochemical cycles.[24,64] The photocatalytic reduction

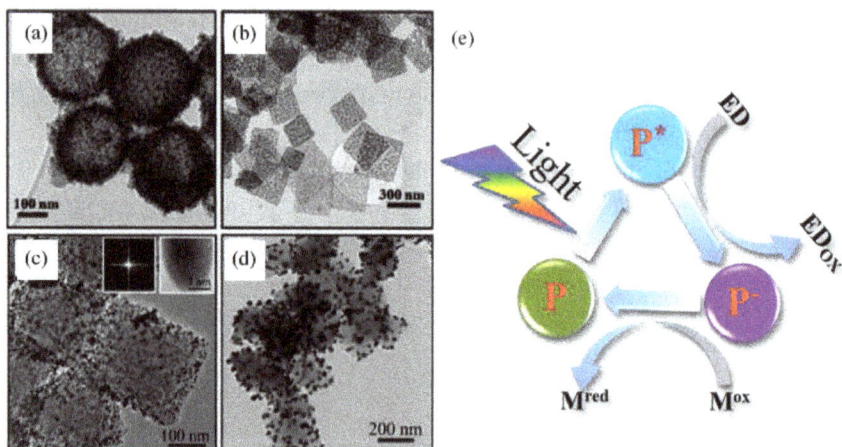

Figure 2.13 Electron microscopy images of the metal nanocrystals grown on tin porphyrin nano-structures. TEM images of (a) Pt/tin porphyrin nanosphere composites; (b) Pt/tin porphyrin nanosheet composites; (c) Pd/tin porphyrin octahedra composites; (d) Ag/tin porphyrin octahedra composites; (e) Photocatalytic reduction mechanism for aqueous metal complexes (for example, Au(I) and Pt(II) complexes) by porphyrin(P), oxidation state metal (Mox) are reduced to reduced state metal (Mred) by the photoinduced electrons (P-), which is transform from photoinduced porphyrin(P*) *via* from electron donor (ED). Adapted with permission from Ref. 35, copyright 2014 American Chemical Society.

mechanism of peptide-porphyrin co-assembly is consistent with the mechanism of one-component porphyrin self-assembly shown in Figure 2.13(e).

2.3.3.2 *Visible-light Photocatalytic Activity for Small Organic Molecule*

The photocatalytic reduction activity of porphyrin assembly has also performed for reduction of some small organic molecule. The aforementioned peptide-porphyrin co-assemblies, prepared by Yan and co-workers, have been also applied to photocatalytic organic species.[24,64] A solution containing 4-nitrophenol (4-NP) and these co-assemblies is irradiated for periods of up to 150 minutes, and detects spectroscopically the time-dependent changes in the concentration of 4-NP and the photoreduced product, 4-aminophenol (4-AP). And the results confirm 4-NP is photocatalytic reduced into 4-AP. The photocatalytic reduction mechanism for 4-NP is also consistent with the mechanism for photocatalytic reduction noble metal salt in Figure 2.13(e).

The specific structure of the porphyrin assembly also performs some stereoselectivity. Stilbene is epoxidized with high photocatalytic activity by self-assembled stacks of trimeric manganese porphyrins (Figure 2.14(a)).[65,66] Small spaces shown in Figure 2.14(b), located between the porphyrin molecules and combined with

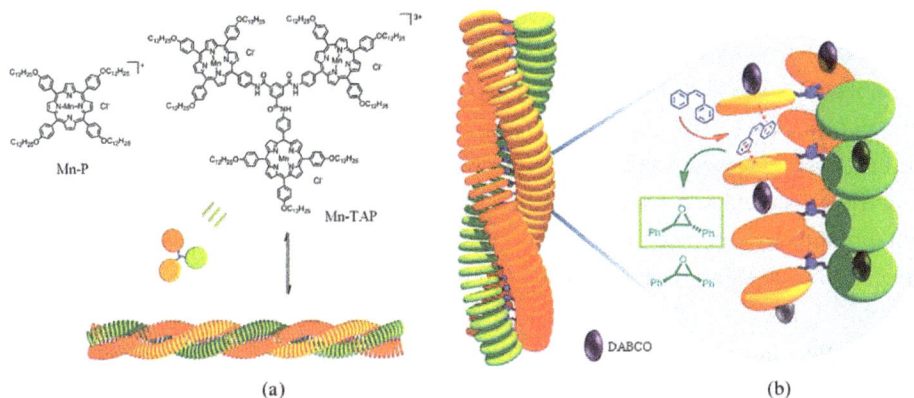

Figure 2.14 (a) Chemical structures of catalysts Mn-P and Mn-TAP and a schematic representation of the proposed columnar stack of Mn-TAP. (b) Proposed schematic model of part of a catalytic stack of Mn-TAP, including DABCO axial ligands, showing the way cis-stilbene enters the stacks to react in the spaces between the porphyrins. Adapted with permission from Ref. 65, copyright 2014 Royal Society of Chemistry.

activation of the Mn-centers by 1, 4-diaza [2, 2, 2] bicyclooctane (DABCO) ligands, are proposed to advance the stereoselectivity of the reaction towards the trans-epoxide by creating a specific steric environment in the stacks.[65] The assembly nanostructure of the catalyst leads to increased stability of the system and enhances the stereoselectivity of the reaction.

2.4 Summary and Outlooks

Porphyrin assembly is one of the most promising visible-light-harvesting and photocatalytic materials for water splitting, degradation of organic pollutants and others. Great breakthroughs on the design and fabrication of porphyrin assembly-based material and along with greater theoretical understanding on the structure-function mechanism have been made in very recent years. To date, various morphology and nanostructures of porphyrin self-assembly, such as nanotube, nanofiber, nanorod, nanoflower, have been successfully constructed by utilization of some self-assembly protocols, including micelle confined self-assembly, micro-emulsion self-assembly, surfactant-assisted self-assembly, ionic self-assembly, metal coordination-assisted self-assembly. Although some specific morphology porphyrin self-assemblies have been controllably synthesized, it is still a big challenge that building regulate and control mechanism that achieve real controllable synthesis for porphyrin assembly-based nanomaterials, because self-assembly process is a spontaneous process through non-covalent bonds and weak interactions such as van der Waals force, hydrogen bond, electrostatic interaction and metal-coordination bonding. Moreover, another big challenge of porphyrin assembly-based visible-light photocatalysts is revealing the mechanism of visible-light photocatalytic reaction. So far, most porphyrin self-assembly researches overemphasize the synthetic strategies and macroscopically visible-light catalytic properties. And only a few researches pay attention to the fundamental visible-light photocatalytic mechanism and establish primary dependency relationship between visible-light photocatalytic activities and morphologies of porphyrin self-assembly. It is expected that, by using a combination of controllable strategies and visible-light photocatalytic mechanism, the visible-light/solar energy conversion efficiency of porphyrin assembly-based materials can be promoted to a level suitable for efficient water splitting and degradation organic pollutants.

References

1. Chen X., Li C., Gratzel M., Kostecki R. and Mao S. S. (2012). Nanomaterials for renewable energy production and storage, *Chem. Soc. Rev.*, 41, 7909–7937.
2. Faunce T., Styring S., Wasielewski M. R., Brudvig G. W., Rutherford A. W., Messinger J., Lee A. F., Hill C. L., deGroot H., Fontecave M., MacFarlane D. R., Hankamer B., Nocera D. G., Tiede D. M., Dau H., Hillier W., Wang L. and Amal R. (2013). Artificial photosynthesis as a frontier technology for energy sustainability, *Energy Environ. Sci.*, 6, 1074–1076.
3. Xia Z., Zhou X., Li J. and Qu Y. (2015). Protection strategy for improved catalytic stability of silicon photoanodes for water oxidation, *Sci. Bull.*, 60, 1395–1402.
4. Yang X., Liu R., He Y., Thorne J., Zheng Z. and Wang D. (2015). Enabling practical electrocatalyst-assisted photoelectron-chemical water splitting with earth abundant materials, *Nano Res.*, 8, 56–81.
5. Sainna M. A., Kumar S., Kumar D., Fornarini S., Crestoni M. E. and de Visser S. P. (2015). A comprehensive test set of epoxidation rate constants for iron (IV)-oxo porphyrin cation radical complexes, *Chem. Sci.*, 6, 1516–1529.
6. Zhou T., Wang D., Chun-Kiat Goh S., Hong J., Han J., Mao J. and Xu R. (2015). Bio-inspired organic cobalt (II) phosphonates toward water oxidation, *Energy Environ. Sci.*, 8, 526–534.
7. Würthner F., Kaiser T. E. and Saha-Möller C. R. (2011). J-aggregates: From serendipitous discovery to supramolecular engineering of functional dye materials, *Angew. Chem. Int. Ed. Engl.*, 50, 3376–3410.
8. Wang T., Chen S. R., Jin F., Cai J. H., Cui L. Y., Zheng Y. M., Wang J. X., Song Y. L. and Jiang L. (2015). Droplet-assisted fabrication of colloidal crystals from flower-shaped porphyrin Janus particles, *Chem. Commun.*, 51, 1367–1370.
9. Huang C. C., Parasuraman P. S., Tsai H. C., Jhu J. J. and Imae T. (2014). Synthesis and characterization of porphyrin-TiO_2 core-shell nanoparticles as visible light photocatalyst, *RSC Adv.*, 4, 6540–6544.

10. Wang Z., Medforth C. J. and Shelnutt J. A. (2004). Porphyrin nanotubes by ionic self-assembly, *J. Am. Chem. Soc.*, 126, 15954–15955.

11. Jin S., Son H.-J., Farha O. K., Wiederrecht G. P. and Hupp J. T. (2013). Energy transfer from quantum dots to metal-organic frameworks for enhanced light harvesting, *J. Am. Chem. Soc.*, 135, 955–958.

12. Lu J., Zhang Y., Jiao C., Megarajan S. K., Gu D., Yang G., Jiang H., Jia C. and Schüth F. (2015). Effect of reduction-oxidation treatment on structure and catalytic properties of ordered mesoporous Cu-Mg-Al composite oxides, *Sci. Bull.*, 60, 1108–1113.

13. Hasobe T., Sandanayaka A. S. D., Wada T. and Araki Y. (2008). Fullerene-encapsulated porphyrin hexagonal nanorods. An anisotropic donor-acceptor composite for efficient photoinduced electron transfer and light energy conversion, *Chem. Commun.*, 3372–3374.

14. Philp D. and Stoddart J. F. (1996). Self-assembly in natural and unnatural systems, *Angew. Chem. Int. Ed. Engl.*, 35, 1154–1196.

15. Whitesides G. M. and Boncheva M. (2002). Beyond molecules: Self-assembly of mesoscopic and macroscopic components, *Proc. Natl. Acad. Sci.*, 99, 4769–4774.

16. Vasilopoulou M., Douvas A., Georgiadou D., Constantoudis V., Davazoglou D., Kennou S., Palilis L., Daphnomili D., Coutsolelos A. and Argitis P. (2014). Large work function shift of organic semiconductors inducing enhanced interfacial electron transfer in organic optoelectronics enabled by porphyrin aggregated nanostructures, *Nano Res.*, 7, 679–693.

17. Wang Y., Shi R., Lin J. and Zhu Y. (2011). Enhancement of photocurrent and photocatalytic activity of ZnO hybridized with graphite-like C_3N_4, *Energy Environ, Sci.*, 4, 2922–2929.

18. So M. C., Beyzavi M. H., Sawhney R., Shekhah O., Eddaoudi M., Al-Juaid S. S., Hupp J. T. and Farha O. K. (2015). Post-assembly transformations of porphyrin-containing metal-organic framework (MOF) films fabricated *via* automated layer-by-layer coordination, *Chem. Commun.*, 51, 85–88.

19. Gong X., Milic T., Xu C., Batteas J. D. and Drain C. M. (2002). Preparation and characterization of porphyrin nanoparticles, *J. Am. Chem. Soc.*, 124, 14290–14291.

20. Sendt K., Johnston L. A., Hough W. A., Crossley M. J., Hush N. S. and Reimers J. R. (2002). Switchable electronic coupling in model oligoporphyrin molecular wires examined through the measurement and assignment of electronic absorption spectra, *J. Am. Chem. Soc.*, 124, 9299–9309.

21. Wang Z. C., Li Z. Y., Medforth C. J. and Shelnutt J. A. (2007). Self-assembly and self-metallization of porphyrin nanosheets, *J. Am. Chem. Soc.*, 129, 2440–2441.

22. Caselli M. (2015). Porphyrin-based electrostatically self-assembled multilayers as fluorescent probes for mercury (II) ions: A study of the adsorption kinetics of metal ions on ultrathin films for sensing applications, *RSC Adv.*, 5, 1350–1358.

23. Jana A., Gobeze H. B., Ishida M., Mori T., Ariga K., Hill J. P. and D'Souza F. (2015). Breaking aggregation in a tetrathiafulvalene-fused zinc porphyrin by metal-ligand coordination to form a donor-acceptor hybrid for ultrafast charge separation and charge stabilization, *Dalton Trans.*, 44, 359–367.

24. Liu K., Xing R., Chen C., Shen G., Yan L., Zou Q., Ma G., Möhwald H. and Yan X. (2015). Peptide-induced hierarchical long-range order and photocatalytic activity of porphyrin assemblies, *Angew. Chem. Int. Ed. Engl.*, 54, 500–505.

25. Wang Q., Chen H., Liu G. and Wang L. (2015). Control of organic-inorganic halide perovskites in solid-state solar cells: A perspective, *Sci. Bull.*, 60, 405–418.

26. Wei P., Yan X. and Huang F. (2015). Supramolecular polymers constructed by orthogonal self-assembly based on host-guest and metal-ligand interactions, *Chem. Soc. Rev.*, 44, 815–832.

27. Zhao Q., Wang Y., Qiao Y., Wang X., Guo X., Yan Y. and Huang J. (2014). Conductive porphyrin helix from ternary self-assembly systems, *Chem. Commun.*, 50, 13537–13539.

28. Qiu Y., Chen P. and Liu M. (2010). Evolution of various porphyrin nanostructures *via* an oil/aqueous medium: Controlled self-assembly, further organization, and supramolecular chirality, *J. Am. Chem. Soc.*, 132, 9644–9652.

29. Zhong Y., Wang J., Zhang R., Wei W., Wang H., Lü X., Bai F., Wu H., Haddad R. and Fan H. (2014). Morphology-controlled self-assembly and synthesis of photocatalytic nanocrystals, *Nano Lett.*, 14, 7175–7179.

30. Guo P. P., Chen P. L., Ma W. H. and Liu M. H. (2012). Morphology-dependent supramolecular photocatalytic performance of porphyrin nanoassemblies: From molecule to artificial supramolecular nanoantenna, *J. Mater. Chem.*, 22, 20243–20249.

31. Hu J.-S., Guo, Liang H.-P., Wan L.-J. and Jiang L. (2005). Three-dimensional self-organization of supramolecular self-assembled porphyrin hollow hexagonal nanoprisms, *J. Am. Chem. Soc.*, 127, 17090–17095.

32. Guo P., Zhao G., Chen P., Lei B., Jiang L., Zhang H., Hu W. and Liu M. (2014). Porphyrin nanoassemblies *via* surfactant-assisted assembly and single nanofiber nano-electronic sensors for high-performance H_2O_2 vapor sensing, *ACS Nano* 8, 3402–3411.

35. Zhong Y., Wang Z., Zhang R., Bai F., Wu H., Haddad R. and Fan H. (2014). Interfacial self-assembly driven formation of hierarchically structured nanocrystals with photocatalytic activity, *ACS Nano.*, 8, 827–833.

36. Ren X. B., Chen M. and Qian D. J. (2012). Pd (II)-mediated triad multilayers with zinc tetrapyridylporphyrin and pyridine-functionalized nano-TiO_2 as linkers: Assembly, characterization, and photocatalytic properties, *Langmuir*, 28, 7711–7719.

37. Sandanayaka A. S. D., Murakami T. and Hasobe T. (2009). Preparation and photophysical and photoelectrochemical properties of supramolecular porphyrin nanorods structurally controlled by encapsulated fullerene derivatives, *J. Phys. Chem. C*, 113, 18369–18378.

38. Bai F., Sun Z., Wu H., Haddad R. E., Coker E. N., Huang J. Y., Rodriguez M. A. and Fan H. (2011). Porous one-dimensional nanostructures through confined cooperative self-assembly, *Nano Lett.*, 11, 5196–5200.

39. Bai F., Wu H., Haddad R. E., Sun Z., Schmitt S. K., Skocypec V. R. and Fan H. (2010). Monodisperse porous nanodiscs with fluorescent and crystalline wall structure, *Chem. Commun.*, 46, 4941–4943.

40. Chen W., Peng Q. and Li Y. (2008). Alq3 nanorods: Promising building blocks for optical devices, *Adv. Mater.*, 20, 2747–2750.

41. Bai F., Sun Z., Wu H., Haddad R. E., Xiao X. and Fan H. (2011). Templated photocatalytic synthesis of well-defined platinum hollow nanostructures with enhanced catalytic performance for methanol oxidation, *Nano Lett.*, 11, 3759–3762.

42. Bai F., Wang D., Huo Z., Chen W., Liu L., Liang X., Chen C., Wang X., Peng Q. and Li Y. (2007). A versatile bottom-up assembly approach to colloidal spheres from nanocrystals, *Angew. Chem. Int. Ed. Engl.*, 46, 6650–6653.

43. Fan H., Chen Z., Brinker C. J., Clawson J. and Alam T. (2005). Synthesis of organosilane functionalized nanocrystal micelles and their self-assembly, *J. Am. Chem. Soc.* 127, 13746–13747.

44. Fan H., Leve E., Gabaldon J., Wright A., Haddad R. E. and Brinker C. J. (2005). Ordered two- and three-dimensional arrays self-assembled from water-soluble nanocrystal-micelles, *Adv. Mater.*, 17, 2587–2590.

45. Gao Y., Zhang X., Ma C., Li X. and Jiang J. (2008). Morphology-controlled self-assembled nanostructures of 5,15-di[4-(5acetylsulfanylpentyloxy)phenyl]porphyrin derivatives. Effect of metal-ligand coordination bonding on tuning the intermolecular interaction, *J. Am. Chem. Soc.*, 130, 17044–17052.

46. Sakuma T., Sakai H. and Hasobe T. (2012). Preparation and structural control of metal coordination-assisted supramolecular architectures of porphyrins. Nanocubes to microrods, *Chem. Commun.*, 48, 4441–4443.

47. Zhou Y., Yang W., Qin M. and Zhao H. (2016). Self-assembly of metal–organic framework thin films containing metalloporphyrin and their photocatalytic activity under visible light, *Appl. Organomet. Chem.*, 30, 188–192.

48. Kc C. B. and D'Souza F. (2016). Design and photochemical study of supramolecular donor–acceptor systems assembled *via* metal–ligand axial coordination, *Coord. Chem. Rev.*, 322, 104–141.

49. Ma Y., Wang X., Jia Y., Chen X., Han H. and Li C. (2014). Titanium based nanomaterials for photocatalytic fuel generations, *Chem. Rev.*, 114, 9987–10043.

50. Chen Y. Z., Li A. X., Huang Z. H., Wang L. N. and Kang F. Y. (2016). Porphyrin-based nanostructures for photocatalytic applications, *Nanomaterials*, 6, 51.

51. Mandal S., Nayak S. K., Mallampalli S. and Patra A. (2014). Surfactant-assisted porphyrin based hierarchical nano/micro assemblies and their efficient photocatalytic behavior, *ACS Appl. Mater. Inter.*, 6, 130–136.

52. Chen C., Ma W. and Zhao J. (2010). Semiconductor-mediated photodegradation of pollutants under visible-light irradiation, *Chem. Soc. Rev.*, 39, 4206–4219.

53. Satake A. and Kobuke Y. (2007). Artificial photosynthetic systems: Assemblies of slipped cofacial porphyrins and phthalocyanines showing strong electronic coupling, *Org. Biomol. Chem.*, 5, 1679–1691.

54. Fujishima A. and Honda K. (1972). Electrochemical photolysis of water at a semiconductor electrode, *Nature*, 238, 37–38.

55. Wang Y., Wang Q., Zhan X., Wang F., Safdar M. and He J. (2013). Visible light driven type II heterostructures and their enhanced photocatalysis properties: A review, *Nanoscale*, 5, 8326–8339.

56. Gan J., Lu X. and Tong Y. (2014). Towards highly efficient photoanodes: Boosting sunlight-driven semiconductor nanomaterials for water oxidation, *Nanoscale*, 6, 7142–7164.

57. Hasobe T., Sakai H., Mase K., Ohkubo K. and Fukuzumi S. (2013). Remarkable enhancement of photocatalytic hydrogen evolution efficiency utilizing an internal cavity of supramolecular porphyrin hexagonal nanocylinders under visible-light irradiation, *J. Phys. Chem. C.*, 117, 4441–4449.

58. Zhu M. S., Du Y. K., Yang P. and Wang X. M. (2013). Donor-acceptor porphyrin functionalized Pt nano-assemblies for artificial photosynthesis: A simple and efficient homogeneous photocatalytic hydrogen production system, *Catal. Sci. Technol.*, 3, 2295–2302.

59. Arai Y., Tsuzuki K. and Segawa H. (2012). Homogeneously mixed porphyrin J-aggregates with rod-shaped nanostructures *via* zwitterionic self-assembly, *Phys. Chem. Chem. Phys.*, 14, 1270–1276.

60. Zhu M. S., Lu Y. T., Du Y. K., Li J. A., Wang X. M. and Yang P. (2011). Photocatalytic hydrogen evolution without an electron mediator using a porphyrin-pyrene conjugate functionalized Pt nanocomposite as a photocatalyst, *Int. J. Hydrogen Energy*, 36, 4298–4304.

61. Morawski O., Izdebska K., Karpiuk E., Nowacki J., Suchocki A. and Sobolewski A. L. (2014). Photoinduced water splitting with oxotitanium tetraphenylporphyrin, *Phys. Chem. Chem. Phys.*, 16, 15256–15262.

62. Sobolewski A. L. and Domcke W. (2012). Photoinduced water splitting with oxotitanium porphyrin: a computational study, *Phys. Chem. Chem. Phys.*, 14, 12807–12817.

63. Si W., Li J., Li H., Li S., Yin J., Xu H., Guo X., Zhang T. and Song Y. (2013). Light-controlled synthesis of uniform platinum nanodendrites with markedly enhanced electrocatalytic activity, *Nano Res.*, 6, 720–725.

64. Zou Q. L., Zhang L., Yan X. H., Wang A. H., Ma G. H., Li J. B., Mohwald H. and Mann S. (2014). Multifunctional porous microspheres based on peptide-porphyrin hierarchical co-assembly, *Angew. Chem. Int. Ed. Engl.*, 53, 2366–2370.

65. De Torres M., Van Hameren R., Nolte R. J. M., Rowan A. E. and Elemans J. (2013). Photocatalytic oxidation of stilbene by self-assembled stacks of manganese porphyrins, *Chem. Commun.*, 49, 10787–10789.

66. Van Hameren R., Schön P., van Buul A. M., Hoogboom J., Lazarenko S. V., Gerritsen J. W., Engelkamp H., Christianen P. C. M., Heus H. A., Maan J. C., Rasing T., Speller S., Rowan A. E., Elemans J. A. A. W. and Nolte R. J. M. (2006). Macroscopic hierarchical surface patterning of porphyrin trimers *via* self-assembly and dewetting, *Science*, 314, 1433–1436.

Chapter 3

Highly Efficient Fluorescent Carbon Quantum Dots: Synthesis, Properties and Applications

Zaicheng Sun and Hongyou Fan

3.1 Introduction

Over the past few centuries, carbon materials have been a practical material in a general use ranging from coal and carbon fibers for our daily life to carbon nanotube, fullerene and graphene for the cutting edge technology. In 2004, Scrivens and coworkers reported first, a new type of fluorescent carbon material during the separation and purification of single-walled carbon nanotube (SWCNT).[1] Later, Sun *et al.* enhanced the fluorescent emission of carbon dots (CDs) *via* surface passivation of amine, which opens a new avenue for the carbon based materials.[2] After that time, the research on the CDs and their derivatives, graphene quantum dots (GQDs) have attracted more and more attention due to their multiple color emission, super photostability, low toxicity and excellent biocompatibility.[3–7] So-called CDs are commonly used for fluorescent carbogenic materials which are quasi-sphere with the typical diameter less than 10 nm. The CDs possess a graphitic core with an outer shell composed of multiple functional groups such as hydroxyl and carboxyl groups. GQDs are not monolayer in nature, but few to tens of layers of graphene with a size below 30 nm and various functional groups on the surface as well.[8] Both CDs and GQDs (denoted as carbon quantum dots, CQDs) can be easily functionalized due to surface functional groups, exhibit high resistance to the photobleaching due to the chemical inertness of the graphitic core, and possess excellent biocompatibility and low cytotoxicity compared with traditional inorganic quantum dots.[9,10] However, they are still suffering from the low photoluminescence (PL)

81

quantum yields (QYs) and excitation wavelength within UV region, which limit their applications in the bioimaging and photocatalyst fields. For the above applications, they require that the CQDs possess visible light absorption and visible light emission to obtain maximum performance. Based on these requirements, significant progress has been made to overcome these limitations to tune the optical properties of CQDs. This chapter focuses on the recent progress of doped carbon nanodots (CNDs) with visible light absorption and high fluorescent PL emission under visible light radiation and related applications.

Since 2004, the fluorescent CQDs have been discovered,[1] many efforts and successes have been made such as high PL QY and tunable emission. CQDs can be treated as a semiconductor, possess typical band gap and unique optical properties. From general semiconductor's view, the band gap of semiconductor can be tuned by introducing heteroatom, so-called dopant. Doped CQDs will exhibit tailored optical and electronic properties. With their nature of low toxicity and excellent biocompatibility, many biorelated and energy conversion related applications have been explored. Here, we only show two typical applications, for other applications like biosensor, light-emitting diode (LED) and organic photovoltaic (OPV) related applications, readers can refer to recent reviews.[7,11,12]

3.2 Synthetic Strategies of CQDs

Many synthesis routes have been developed to prepare CQDs within last decade, which are generally classified into two categories: "top-down" and "bottom-up" routes (Figure 3.1).[3,13] So-called "top-down" route refers to breaking down larger

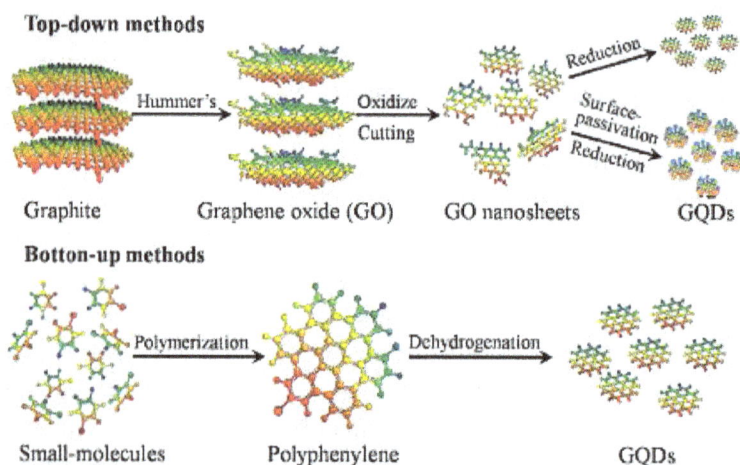

Figure 3.1   Schematic diagram of the top-down and bottom-up methods for synthesizing CQDs. From Ref. 6, *Chem. Commun.*, 2012, 48, 3686–3699. Reprinted with permission from RSC.

carbon structures like graphite,[2,14,15] carbon nanotube,[16,17] carbon fibers,[18] activated carbon,[19] coal[20,21] by numerous physical or chemical means such as arc
discharge, laser ablation, oxidants, electrochemical oxidation, and then passivating as-prepared carbon nanostructures through chemical treatment to enhance
PL emission. Comparing with top-down route, bottom-up route provides a
versatile route to prepare high efficiency CQDs with various compositions.
The carbon source comes from all kinds of natural materials like citric acid,[22,23]
glucose,[24,25] ascorbic acid,[26] chloroform, orange juice,[27] egg white,[28] chitosan,[29]
grass[29] and so on. The different composition of carbon source may bring some
heteroatoms into the CQDs to form doped CQDs with high PL quantum yield
and different emission.

3.2.1 *Synthesis of Non-doped CQDs*

Most CQDs prepared from top-down route are in non-doped state because they
start from graphite carbon materials through oxidation route. The carboxyl,
carbonyl, hydroxyl, or epoxy groups form in the oxidation process. These groups
are contributed to enhance water solubility and functionalization. Although these
carbon nanoparticles can emit broad light, the emission is quite weak. The N passivation process is necessary to improve the PL quantum yield. For example,
Scrivens *et al.* discovered the first example of fluorescent CDs when they were
purifying SWCNTs from arc-discharged soot. The arc-discharged soot was oxidized with nitric acid, extracted using sodium hydroxide solution and the black
suspension from the extract was then subjected to gel electrophoresis to obtain
CQDs.[1] Later, various methods have been employed to cut the graphite carbon
materials to obtain nanosize CQDs. Although these CQDs showed multiple color
emission, the emission peak mainly locates at ~450 nm in blue light region under
the UV radiation. The emissions in the green and red light regions are quite weak
or even hard to be detected. In most cases, the non-doped CQDs show PL QY
below 10%. On the other hand, CQDs can also be prepared *via* bottom-up route.
Wang *et al.* reported that CQDs with blue and green emission were obtained by
hydrothermal treatment of glucose in the presence of monopotassium phosphate.[25]
However, the PL QY of CQDs is about 2.4% and 1.1% for CQDs with blue and
green emissions, respectively. And their PL lifetime decays are near 2.0 ns and
1.8 ns for CQD-Blue and CQDs-Green, respectively. Hydrothermally treated citric
acid in the presence of NaOH formed CQDs with blue light emission with ~20%
of PL QY.[30] Mohapatra *et al.* reported the green fluorescent CQDs by hydrothermal treatment of orange juice at 120°C (Figure 3.2).[27] The CQDs formed by
the carbonization of sucrose, glucose, fructose, citric acid and ascorbic acid in the
orange juice. The PL QY of CQDs can reach as high as 26%.

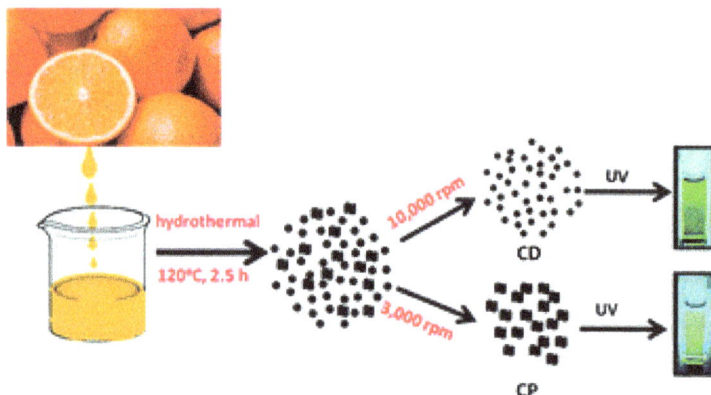

Figure 3.2 Schematic synthesis route of non-doped CQD. From Ref. 27, *Chem. Commun.,* 2012, 48, 8355. Reprinted with permission from RSC.

3.2.2 *Synthesis of N-doped CQDs*

Although CQDs show relatively low fluorescent emission, Sun *et al.* discovered that the PL QY of CQDs can be improved up to 40% by surface passivation with PEG 1500N and above.[2] These results indicate that the PL QY of CQDs can be improved by introducing N atoms to form N-doped CQDs. Giannelis and coworkers took the lead in carbonization of citrate salt and obtained organophilic nanoparticles from octadecyl ammonium and hydrophilic nanoparticles from 2-(2-aminoethoxy)-ethanol salt.[31] Dehydration of citric portion enables to grow uniform carbon or graphite core covalently tethered to the organic corona from the corresponding octadecyl or hydrophilic groups. As-prepared CDs have near spherical morphology with an average size of 7 nm and near excitation-dependent PL emission. The PL quantum yield is about 3% relative to the Rhodamine 6G at excitation wavelength of 495 nm. Further, this group developed new precursors — tris(hydroxymethyl) aminomethane (Tris) and betaine hydrochloride as carbon source and passivation agent to synthesize CDs.[32] The as-obtained CQDs also show a broad absorption from UV to visible light region. The quantum yield was calculated at 4%, relative to anthracene ethanol solution. Carbohydrates like glucose, glucosamine, sucrose, dextran, cellulose and ascorbic acid are typical carbon sources to synthesize carbon or graphite. Jana *et al.* mixed carbohydrate with fatty amine and then heated the mixture in octadecene to 70–300°C for 10–30 minutes to prepare the hydrophobic CQDs with blue or green emission.[33] If the carbohydrate is dissolved in HCl or NaOH aqueous solution, which is heated at 70–90°C for 20–30 minutes, the hydrophilic CQDs with blue or green emission were obtained. The as-obtained CQDs exhibit tunable color emission and relative high fluorescent quantum yield of 6–30% in reference to fluorescein or quinine sulfate.

With citric acid and a series of diamines, like diethylenetriamine (DETA), triethylenetetraamine (TETA), tetraethylenepetaamine (TEPA) and polyethylenepolyamine (PEPA) as passivation agents, high PL quantum yield CDs can be synthesized by one-step thermolysis at 170°C.[34] The CQDs have uniform particle size with average size of 4.0 nm. The fluorescent QY of as-obtained CQDs reach 88.6% in reference to coumarin 1 at excitation wavelength of 360 nm. Hydrothermal is a common route to synthesize high QY fluorescent CQDs, because it effectively forms N-doped CQDs. Yang *et al.* reported that CQDs with near 80% PL QY can be obtained from citric acid and ethylene diamine by this route.[35] Furthermore, the PL QY of CQDs can be improved to 94% by Sun's group.[30] Comparing with other methods, hydrothermal route provides a uniform reaction environment. The dopant amount is also easily tuned by reaction conditions, like N source and reaction temperature and times. The as-prepared N-doped CQDs also show unique excitation-independent emission, which indicates a single emission center in the CQDs. By tracking the reaction of CQD formation (Figure 3.3), citric acid can dehydrate in the base media and form CQDs. Due to the existence of amine, part of carboxyl group was transferred into amide group. Then the N was brought into the graphite structure. The CQDs prepared from hydrothermal route have a high PL lifetime decay of more than 10 ns.

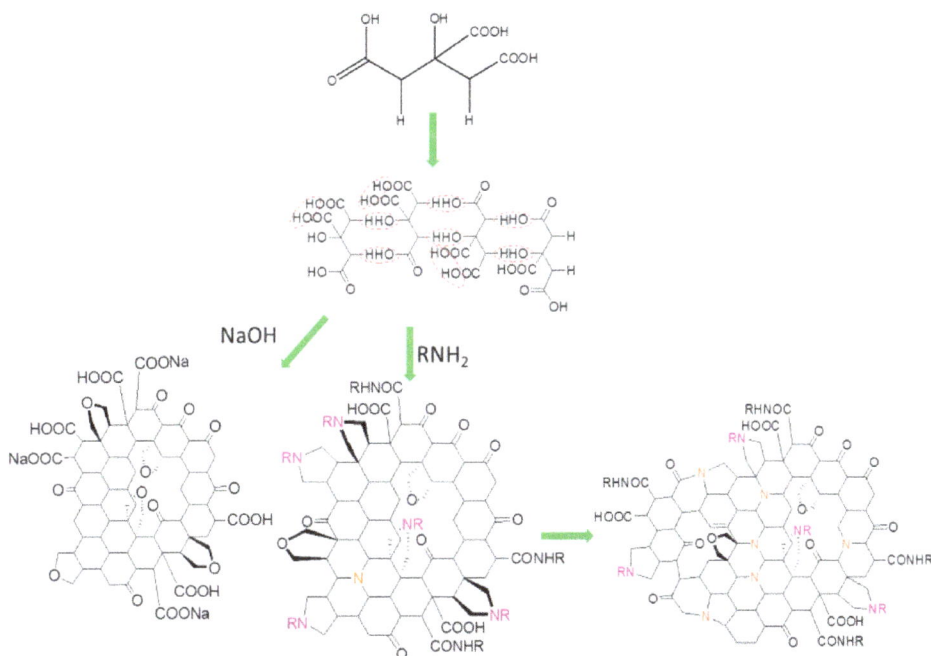

Figure 3.3   The possible reaction route to form N-doped CQDs. From Ref. 30, *Sci. Rep.* 2014, 4, 5294. Reprinted with permission from NPG.

Generally, CQDs prepared in water media exhibit highly efficient blue light emission through hydrothermal route. Recently, Sun *et al.* reported that the PL of N-doped CQDs could be tuned from blue to yellow by changing the reaction conditions. When the reaction solvent changed from water to dimethyl formamide (DMF), the CQDs exhibit green light emission with PL QY of ~29%. Furthermore, the CQDs with yellow light emission can be prepared from citric acid and DETA in absence of any solvent. The PL QY of CQDs with yellow emission also reaches ~22%. The lifetime decays of all these three CQDs are single exponential decays of 14 ns, 13 ns and 10 ns for blue, green and yellow emissions, respectively.[36]

Recently, Lin group reported CQDs with red, green and blue emissions prepared from *p-*, *o-*, *m*-phenyldiamine *via* solvothermal route (Figure 3.4).[37] The N content is 3.69%, 7.32% and 15.57% for N-doped CQDs with the maximum emission at 435 nm, 535 nm and 604 nm, respectively. The PL QY of these CQDs are 4.8, 17.6 and 26.1 for blue, green and red excited at 360 nm, 420 nm and 510 nm, respectively. They also observed up-conversion PL under an 800 nm femtosecond pulsed laser.

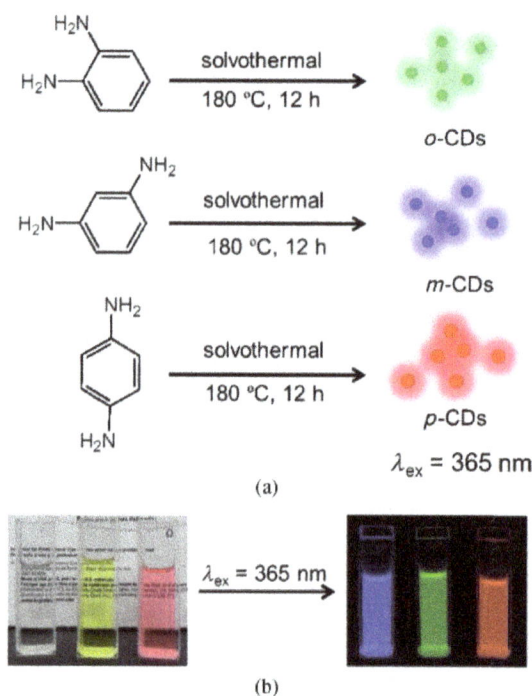

Figure 3.4 (a) preparation of the RGB PL CQDs from three different phenylenediamine isomers (i.e. *o*PD, *m*PD and *p*PD). (b) photographs of *m*-CQDs, *o*-CQDs and *p*-CQDs dispersed in ethanol in daylight (left) and under $\lambda = 365$ nm UV irradiation (right). From Ref. 37, Reprinted with permission from Wiley.

3.2.3 Synthesis of S-doped CQDs

Doping is a common technique for tuning the band gap of semiconductors. Here, the concept of doping is borrowed from semiconductors. The heteroatoms are introduced into the CQDs for tuning the fluorescent intensity or emission position. Li *et al.* reported S-doped CQDs could be obtained by breaking graphite rods in the 0.1 M sodium *p*-toluenesulfonate aqueous solution *via* electrochemical route.[38] X-ray photoelectron spectroscopy (XPS) results reveal that the most of S atoms exist as –C–S–C covalent bond of the thiophene-like S. Other S atoms are present in the form of –C–S(O)$_2$–C– sulphone bridges. The as-prepared S-doped CQDs display a near excitation-independent emission. The maximum excitation and maximum emission wavelengths are at 380 nm and 480 nm, respectively. The QY and lifetime of S-doped CQDs are as high as 10.6% and 5.8 ns, respectively, when excited at 380 nm and emitted at 480 nm. Kang *et al.* prepared S-doped CQDs by hydrothermally treating the oxidized CQDs in the NaSH at 240°C for 12 h.[39] The S-doped CQDs exhibit green excitation-independent luminescence under UV irradiation. The maximum excitation and emission wavelength shift to 418 nm and 539 nm. The QY and lifetime of S-doped CQDs are about 24% and 4.85 ns, respectively. Recently, Xu *et al.* Directly prepared S-doped CQDs from bottom-up route.[40] Sodium citrate and sodium thiosulfate were mixed together and hydrothermally treated at 200°C for 6 h (Figure 3.5). The as-prepared S-doped CQDs exhibit a bright blue emission under UV excitation, which is excitation-independent. The QY and lifetime of S-doped CQDs reach 67% and 11.26 ns, respectively. XPS disclose that the most of S exists in the form of –C–SOx ($x = 2, 3, 4$).

3.2.4 Synthesis of S, N Co-doped CQDs

Citric acid and L-cysteine were used to produce N, S co-doped CQDs through a one-step hydrothermal treatment.[41] The citric acid serves as the carbon source, while the L-cysteine provides nitrogen and sulfur. The as-prepared N, S-CQDs

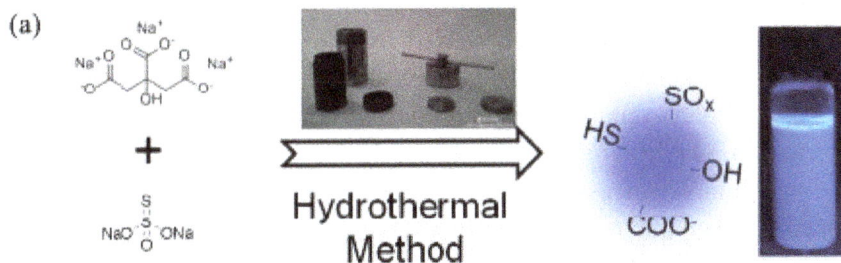

Figure 3.5 Schematic synthesis route of S-doped CQDs from sodium citrate and sodium thiosulfate. From Ref. 40, *J. Mater. Chem. A*, 3, pp. 542–546. Reprinted with permission from RSC.

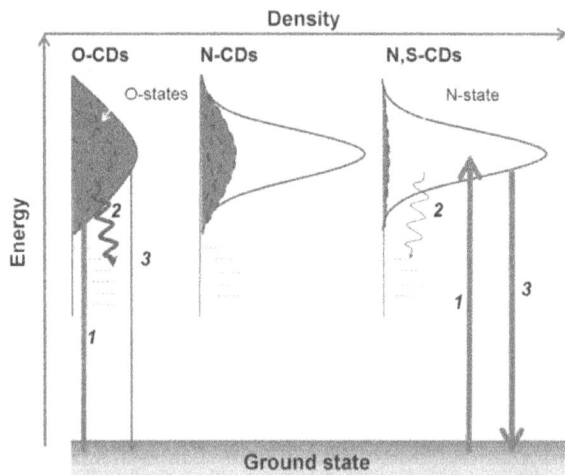

Figure 3.6 Representation for the PL mechanism of O-doped CQDs, N-doped CQDs and S, N co-doped CQDs. (1) electrons excited from the ground state and trapped by the surface states; (2) excited electron return to the ground state *via* a non-radiative route; (3) excited electrons return to the ground state *via* a radiative route. Ref. 41, Reprinted with permission from Wiley.

exhibit very high PL QY (73%), long lifetime (12.11 ns) and excitation-independent emission (435 nm), resulting from the synergy effect of the doped nitrogen and sulfur atoms. The author compared three CQDs (CQDs, N-doped CQDs and S, N co-doped CQDs). The PL mechanism (Figure 3.6) was proposed to explain the PL processes of the three CQDs. The original CQDs have different surface states, denoted as O-states, corresponding to a relatively wide distribution of different energy levels to generate a broad UV–Vis absorption band and broad excitation-dependent emission. The nitrogen doping introduced a new kind of surface state (labeled as the N-State). Electrons trapped by the newly formed surface states are able to facilitate a high yield radiative recombination. The PL spectra of N-doped CQDs are still broad and excitation-dependent. It seems that the introduction of S would enhance the effect of nitrogen on the properties of the doped CQDs, Therefore, N, S-CQD shows high PL QY and excitation-independent emission. Sun *et al.* started from citric acid mixed with urea or thiourea and obtained N and S, N co-doped CQDs.[42] N-doped CQDs exhibit high PL QY of 78% and excitation-independent blue emission at 435 nm with 360 nm UV excitation. The lifetime of N-doped CQDs has a single exponential decay of 7.6 ns, which indicates a single fluorescent origin in the N-doped CQDs. S, N co-doped CQDs exhibit clear absorptions at ~ 550 nm and 595 nm, which are related to the S-doping. The maximum emission of S, N co-doped CQDs are at 450 nm with PL QY and lifetime of 71% and 12.8 ns, respectively. However, if the excitation wavelength shift to longer wavelength it still can emit red light.

Xiong *et al.* investigated S and S, N co-doped CQDs prepared from α-lipoic acid in the base solution (NaOH) with the absence and presence of ethylene diamine.[43] S-doped CQDs exhibit excitation-dependent blue emission with a QY of 5.5%. S, N co-doped CQDs exhibit near excitation-independent blue emission with a PL QY of 54.4%. That indicates that the N-doping plays a critical role for the improvement of PL QY of CQDs.

If water is replaced with *N,N*-dimethyl formamide, S, N co-doped CQDs also can be obtained from citric acid and thiourea.[44] However, S, N co-doped CQDs exhibit different optical properties. For example, there exist two more absorption bands at 467 nm and 557 nm (Figure 3.7(a)). The uniqueness of the S, N co-doped CQD is that it exhibits a single particle with blue, green and red multiple color emissions. The PL QYs of S, N-CQDs are 61%, 45% and 8% for blue, green and red light emissions, respectively. The PL decay of S, N-GQDs could be fitted by a double exponential function (Figure 3.7(e)). The average lifetimes of the S, N-GQDs emission were 10 ns and 6 ns for excitations at 340 nm and 440 nm, respectively.

3.2.5 *Synthesis of P-doped CQDs*

N-doping greatly improves the PL intensity and QY of CQDs, element P has similar electron configuration as element N. Although phosphorus atom is larger than carbon atom, it has been shown that phosphorous can form substitutional defects in diamond sp^3 thin films, behaving as an n-type donor and thereby modifying the electronic and optical properties. P-doped CQDs were synthesized from phosphorous bromide and hydroquinone through hydrothermal reaction.[45] As-prepared P-doped CQDs present strong blue fluorescence with PL QY of 25% and average lifetime of 5.0 ns. It clearly indicates that the P-doping can improve the photoluminescent quantum yield, as well. Liu and coworkers discovered, P-doped CQDs can be easily obtained by one-step acidic oxidation of sucrose by H_3PO_4 at low temperature (60°C).[46] The supernatant contains the CQDs with green emission and the precipitate exhibits yellow fluorescence. The author confirmed the P exists in the C–O–P=O in both green and yellow P-doped CQDs by XPS and ^{31}P-NMR due to thermal dehydration of hydroxyl phosphate moieties between OH on the surface of carbon core and H_3PO_4. The PL QYs of P-doped CQDs are 4% and 3% for green and yellow, respectively. Furthermore, Jana group also demonstrated the carbohydrate can be carbonized to form P-doped CQD with yellow and red emissions by concentrated phosphate acid at 80–90°C.[33] Although these CQDs exhibit similar size, they display different emissions. The main reason is the different surface oxidation degree, which results in the different surface states with energy levels between π and π* state of C=C and induces

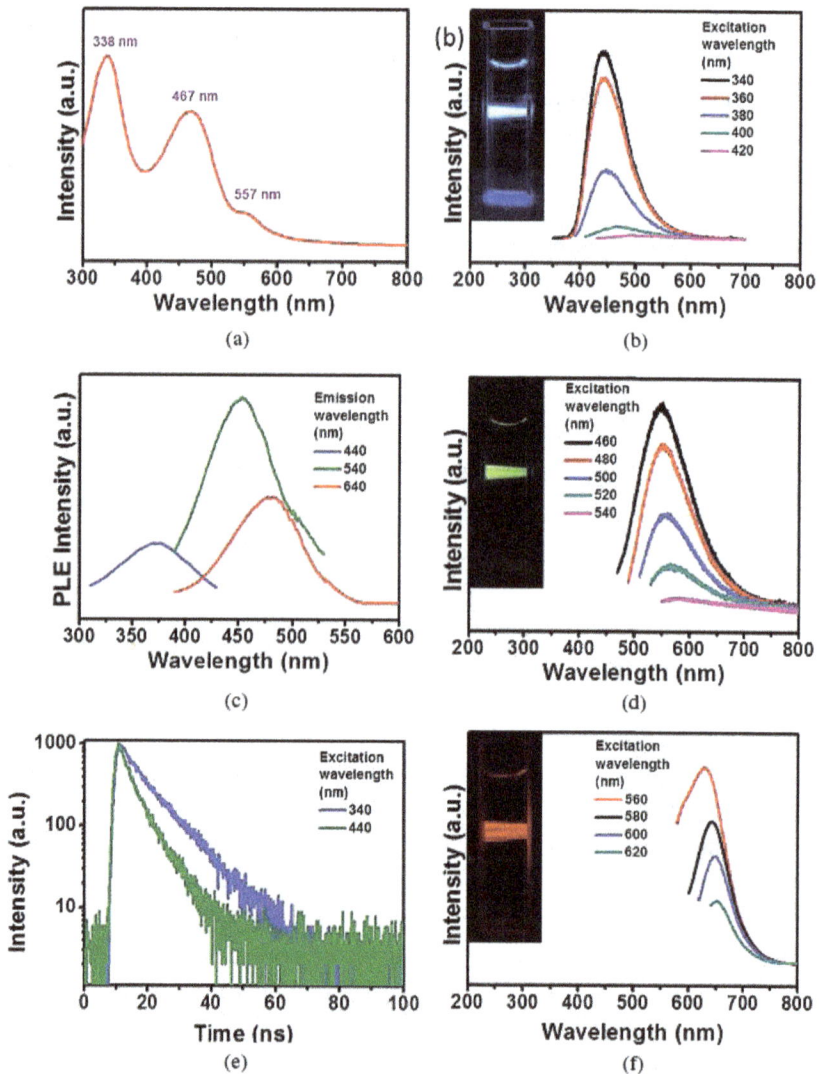

Figure 3.7 (a) UV–Vis adsorption, PL spectra of S, N co-doped CQDs under different excitation wavelengths in the ranges of 340–420 nm (b), 460–540 nm (d), and 560–620 nm (d). The insets in (b), (d), (f) are optical photographs of the S, N co-doped CQDs aqueous excited at 340 nm, 460 nm and 560 nm, respectively. PLE spectra (c) and PL lifetime decay (e) of S, N co-doped CQDs.[44] Reprinted with permission from Wiley.

different absorption bands within these groups. Phytic acid is a phosphorus rich compound, which could be employed as P source for synthesis of CQDs. P-doped CQDs also are prepared *via* microwave route from phytic acid and EDA aqueous (Figure 3.8).[47] The purified P-doped CQDs with acetone exhibit ~20% PL QY and 4.88 ns average lifetime.

(a)   (b)

Figure 3.8 Schematic diagram for synthesizing PCDs. (b) Emission under daylight and 365 nm ultraviolet excitation.[47] Reprinted with permission from RSC.

3.2.6 *Other Dopants Like Si, Cl and Se, B, Gd, ZnS*

3.2.6.1 *Si-doped CQDs*

The doping of silicon (Si), which belongs to the same group as C, group IV, but strongly prefers sp³-like bonding, is considered to influence the electronic and structural properties of carbon-based material in a different way from that of N-doping. Feng *et al.* demonstrated that the Si-doped CQDs can be obtained from SiCl$_4$ and hydroquinone in acetone *via* solvothermal route.[48] Si-doped CQDs showed maximum emission at 382 nm with excitation at 300 nm, PL QY of 19.2% and average lifetime of 3.3 ns.

3.2.6.2 *Cl-doped CQDs*

Cl-CQDs were prepared, using fructose and hydrochloric acid as source materials, under hydrothermal conditions at 170°C over 4 h HCl plays two roles during the preparation of Cl-CQDs.[49] One is to catalyze the dehydration of fructose under acidic conditions. The other is to provide a Cl-doping source. The emission color of the Cl-CQDs can be tuned successively from blue to white, orange, green and red simply by varying the excitation wavelength between 300 nm and 600 nm (Figure 3.9). The multicolor emission from the Cl-CQDs can be attributed to Cl-doping, which introduces additional energy levels between C π and C π^*.

3.2.6.3 *Se-doped CQDs*

Se locates in the same group as S and O, but has less electronegativity (Φ_f: 2.48). Kang *et al.* prepared Se-doped CQDs from oxidized CQDs and NaSeH.[39] The Se-doped CQDs exhibit yellow emission at 563 nm with excitation wavelength of 408 nm. The PL QY and average lifetime of Se- doped CQDs are 19% and 3.27 ns, respectively. As shown in Figure 3.10, it is clear that the λ_{em} of heavy

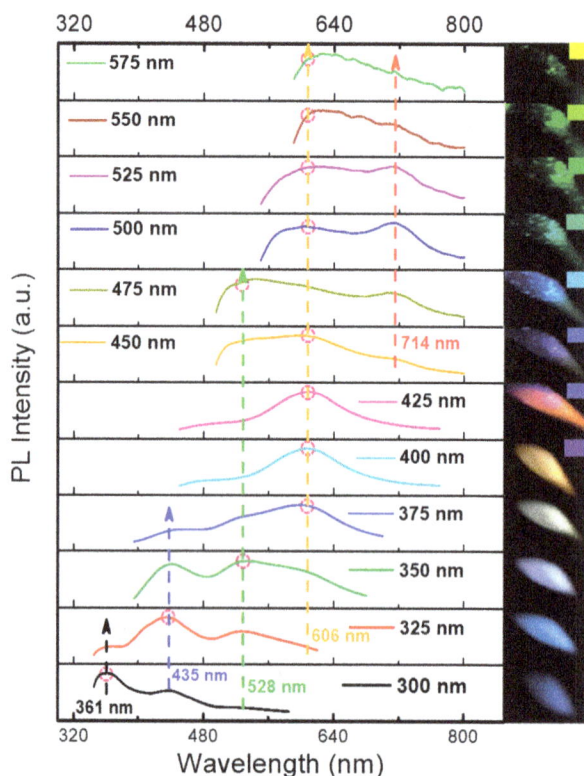

Figure 3.9   The emission spectra of Cl-doped CQDs solution excited by wavelength ranging from 300 nm to 575 nm. The corresponding photographs of the Cl-doped CQDs solution excited by different wavelengths are shown on the right-hand side of the spectra. The inset in the top-right corner of some of the panels shows the color corresponding to the excitation wavelength.[49] Reprinted with permission from RSC.

doped CQDs is related to the electronegativity (χ) of heteroatoms. When the CQDs were doped by S or Se, the fluorescence spectrum red shifted. This can be attributed to the low electronegativity of S and Se, which can be an electron donor. Furthermore, the Φ_f of heavy doped CQDs is also related to the χ of heteroatoms. The results indicate that CQDs possess good conjugation. When linked with the electron acceptor (N atom), the charge transfer is obvious, thus showing high Φ_f. Otherwise, the charge transfer is not obvious, the Φ_f is lower than that of N-CQDs. It is noteworthy that although O is a strong electron acceptor, the oxidized CQDs also show weak luminescence. This is due to lots of oxygen-containing groups (such as –COOH and epoxy), always inducing non-radiative recombination of localized electron-hole pairs and holding back the intrinsic state emission. Similar method was employed by Ding's group.[50] They started from graphene oxide quantum dots (GOQD) prepared from Hummer's route and hydrothermally treated GOQD with NaSeH at 250°C for 24 h. Se-doped CQDs showed yellow emission at 563 nm with 29% PL QY and 3.44 ns average lifetime.

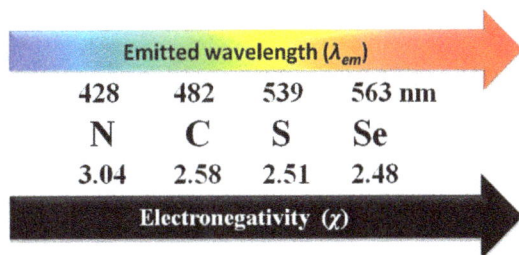

Figure 3.10 The relationship between the electronegativity (χ) of the heteroatom and the λ_{em} of doped CQDs.[39] Reprinted with permission from RSC.

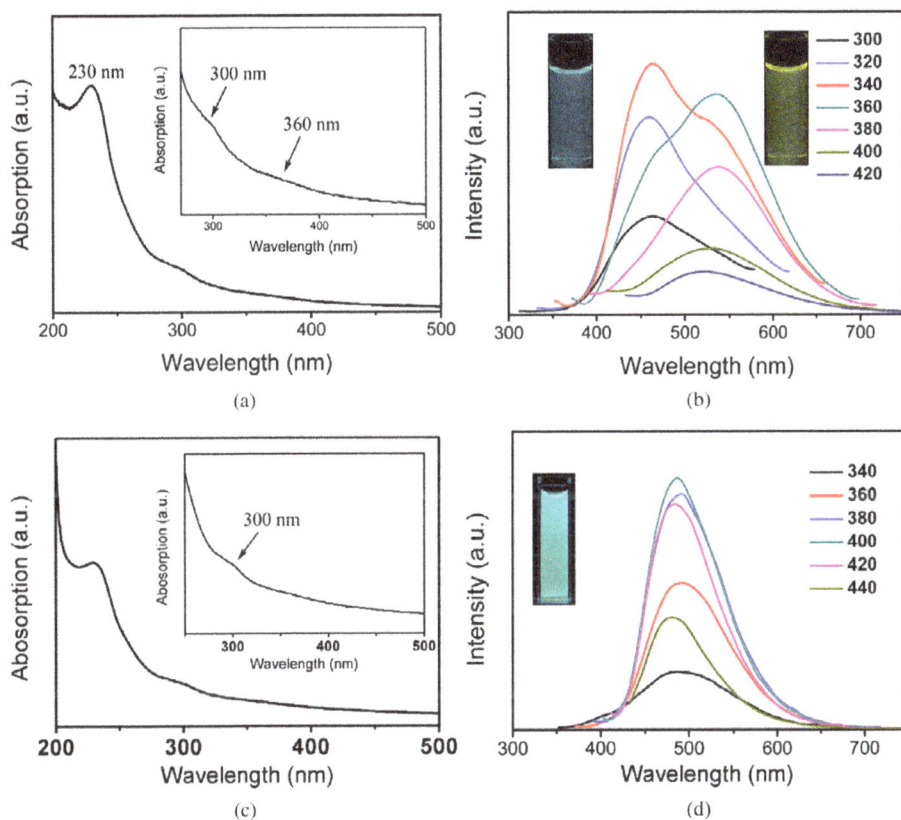

Figure 3.11 UV–Vis absorption and PL spectra of the B-doped CQDs aqueous in the absence (a, b) and presence (c, d) of borax. The insets in (b) and (d) show photographs of the B-doped CQDs solution under 254 and 365 nm UV illumination, respectively.[51] Reprinted with permission from Elsevier.

3.2.6.4 *B-doped CQDs*

In the periodic table, B is located to the left of C and has less electronegativity. The electrochemical preparation of B-CQDs was performed in 0.1 M borax aqueous solution with graphite rod as anode and Pt foil as cathode.[51] B-doped CQDs

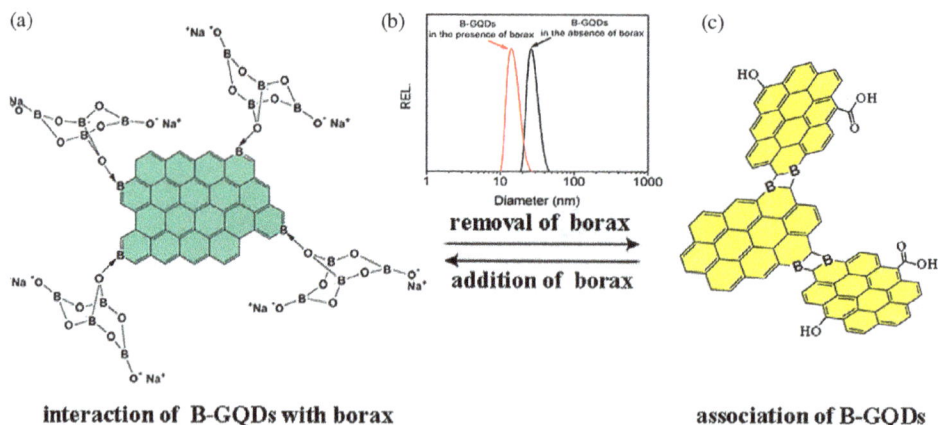

Figure 3.12 The size distribution of B-doped CQDs in the presence (left) and absence (right) of borax determined by DLS (b). the structure illustration of B-doped CQDs in presence (a) and absence (c) of borax, respectively.[51] Reprinted with permission from Elsevier.

exhibit unique PL properties, that is a single particle emits different color lights (Figure 3.11). When the excitation wavelength was set at 300–340 nm, the emission at 460 nm takes the dominating role. When the excitation wavelength shifts to 360–420 nm, the emission at 535 nm becomes the maximum. This phenomenon has been confirmed by recent reports.[44] However, B-doped CQDs only emit one fluorescence in the borax media (Figure 3.12). The PL QY also rise from ~3% up to 13% for the absence and presence of borax. The unique optoelectronic properties of B-CQDs should be associated with the doping of B into the CQDs structures. The bright green fluorescence arises from the B-doping as well as the interaction between B-CQDs and borax. Since B^{3+} is bonded to the sp^2 clusters of CQDs, the short distance and energy level overlap could lead to effective energy transfer from B^{3+} to the sp^2 cluster. As a result, the radiative recombination rate is increased and the PL emissions are consequently enhanced. On the other hand, the special electron deficiency of B atom gives rise to the interaction between B-doped CQDs and O atom of $Na_2B_4O_7$ nearby, which could also be responsible for the bright green fluorescence. The boronic functionalized C-dots are obtained by one-step hydrothermal carbonization, using phenylboronic acid as the sole precursor.[52] It shows strong emission at 400 nm with the UV light excitation.

3.2.6.5 *Gd-doped CQDs*

Rare earth elements can also be employed to dope CQDs. Besides non-metal elements are extensively used for tuning the fluorescent properties of CQDs. Giannelis group introduced the gadopentetic acid into the mixture of

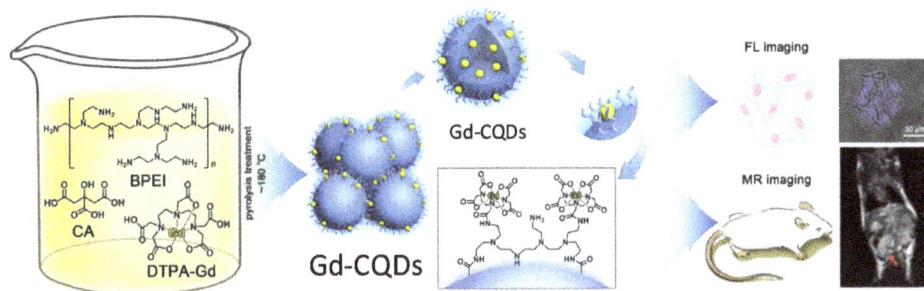

Figure 3.13 Schematic illustration for the synthesis and dual-modality imaging application of Gd-doped CQDs (Gd-CQDs).[55] Reprinted with permission from RSC.

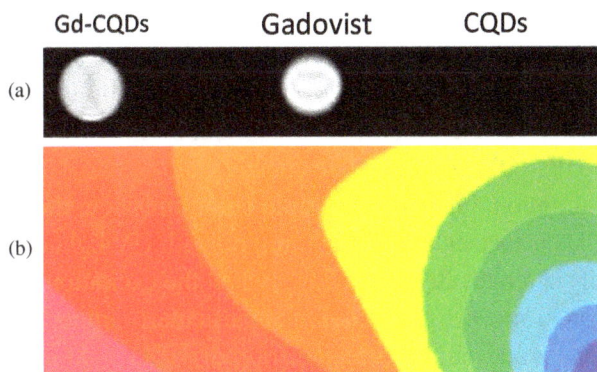

Figure 3.14 (a) MRI positive contrast effects in T1-weighted images of Gd- CQDs, commercial Gd-based contrast agent — Gadovist (both samples were diluted in water on the same concentration of Gd), and un-doped CQDs. All images were obtained using a 1.5 T clinical MRI tomograph. (b) a brown aqueous dispersion of Gd-QCDs (bottom) displays bright purple and green emissions under a fluorescence microscope (the excitation wavelengths were 360–370 nm for purple emission and 460–495 nm for green emission).[53] Reprinted with permission from RSC.

tris(hydroxymethyl) aminomethane and betaine hydrochloride and then heated the mixture in air at 250°C.[53] After carbonization, the Gd-doped CQDs water dispersion exhibits typical excitation-dependent multiple color emission (Figure 3.14(b)). On the other hand, Gd compound shows a good T1 contrast enhancement, which endows CQDs with magnetic property. Gd-doped CQDs exhibit not only fluorescence but also strong T1-weighted MRI contrast (Figure 3.14(a)). These features make them as promising candidates for biomedical application. Following this idea, Gd-doped green CQDs were also prepared from sucrose and Gd^{3+} *via* microwave route.[54] The PL QY of Gd-doped CQDs is about 5.4%. Meanwhile, the r_1 relaxivity of Gd-CDs was measured to be 11.356 mM^{-1} s^{-1}. This high r_1 value together with the r_2/r_1 ratio close to 1 nominates Gd-CDs as an excellent T_1

contrast agent for magnetic resonance imaging. Very recently, Yi *et al.* reported a facile synthesis route to obtain Gd-doped CQDs using citric acid, branch polyethylene imine and Gd-DTPA complexes as source (Figure 3.13).[55] The Gd-CQDs displayed are observed to have a high MR response with longitudinal relaxation of 57.42 $mM^{-1}s^{-1}$ and a strong fluorescent emission with 40% PL QY while containing only 1.0 wt. % Gd^{3+} content. Figure 3.14 shows MRI positive image and Fluorescent images of Gd-doped CQD. Gd-doped CQDs exhibit equivalent MRI signal as commercial Gd contrast agent and multiple color emission.

3.3 Optical Properties of CQDs

CQDs are very attractive fluorescent materials with potential to overcome the disadvantages of organic dyes and inorganic quantum dots due to their good photostability, excitation-dependent emission, low cytotoxicity and good biocompatibility. However, they have their own shortcomings. PL QY of CQDs is an important parameter for their fluorescent application like bioimaging. However, initially reported PL QY of CQDs are mostly below than 10%.[3] Although the CQDs exhibit excitation-dependent multiple color emission, the maximum emission of most CQDs is located at blue light region with UV light excitation, and the emission intensity dramatically drops down when the excitation wavelength shifts to red light region. The red light emission is particularly important for the bioimaging applications since red and near infra-red (NIR) light give deeper tissue penetration. To detect the fluorescent probe in deeper tissue, the excitation wavelength is required to shift to red light region. On the other hand, CQDs can be employed as a sensitizer for the visible light absorption of photocatalyst TiO_2. From the UV–Vis absorption spectra, the absorption of CQDs needs improve further to absorb more visible light.

In most cases, CQDs exhibit a strong broad optical absorption in the UV region with a tail extending into the visible region. The PL spectra are typically broad and dependent on the excitation wavelength, the maximum intensity of PL peaks gradually shifted to the longer wavelength as the excitation wavelength was from 300 nm to 600 nm. The strongest peak was blue light emission at ~450 nm, which was excited at 360 nm (Figure). The PL lifetime of CQDs also showed multiple exponential decay, which means multiple emission centers exist in the CQDs. The N-doped CQDs synthesized from citric acid and ethylenediamine *via* hydrothermal route display different optical properties. There exist two clear optical absorption bands at 235 nm and 340 nm which are assigned to $\pi \to \pi^*$ transition of C=C of sp^2 C domain in sp^3 C matrix and $n \to \pi^*$ transition of conjugated C=O. N mainly exists in form of pyrrolic N and graphite in the N-doped CQDs. The PL spectra show only emission at ~450 nm (blue emission), which showed excitation-independent

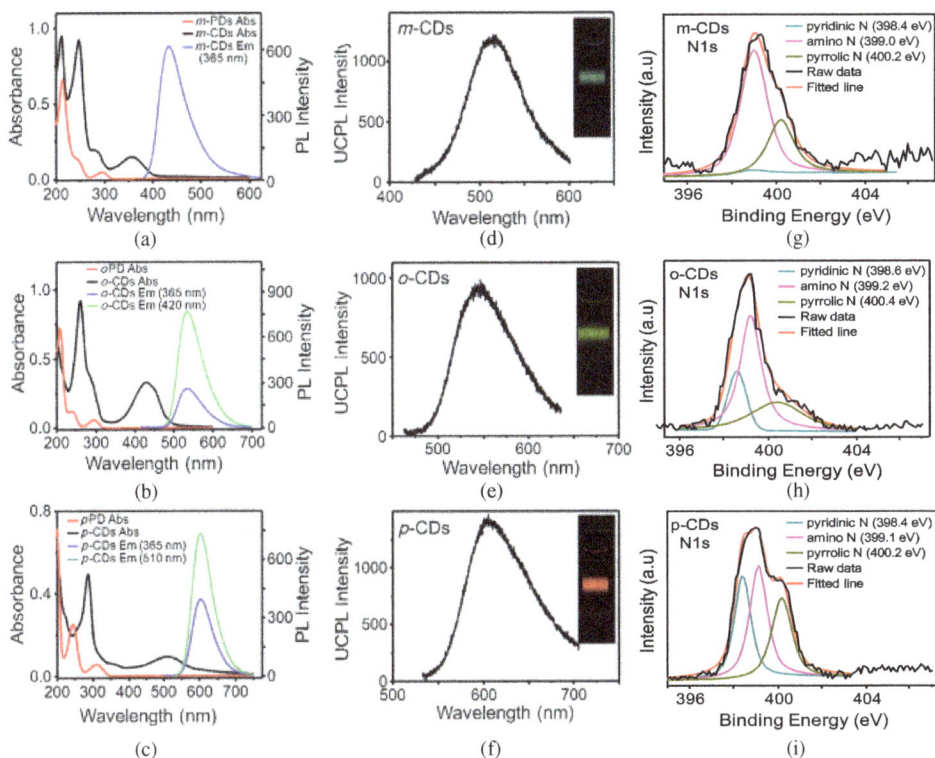

Figure 3.15 (a)–(c) UV–Vis absorption spectra of *m*PD, *o*PD, and *p*PD (red line), *m*-CQDs, *o*-CQDs, and *p*-CQDs (black line), and PL emission spectra of *m*-CQDs, *o*-CQDs, and *p*-CQDs (blue and green lines). (d)–(f) UCPL spectra and photographs (insets) of *m*-CQDs, *o*-CQDs, and *p*-CQDs in ethanol under a λ = 800 nm pulsed laser excitation. Deconvoluted high-resolution XPS spectra of *m*-CQDs (g), *o*-CQDs (h), and *p*-CQDs (i) for N1s.[37] Reprinted with permission from Wiley.

emission. The PL QY of N-doped CQDs can reach as high as 94%. It seems N surface state induced by N-doping facilitates a high yield of radiate recombination. If the solvent changes from water to DMF or there is no solvent, the n → π* transition of conjugated C=O has a slight red-shift due to the large effective π conjugation length. The absorption band at 340 nm shifts to 350 nm. The PL excitation and emission spectra have dramatical shifts to 440 nm and 550 nm for green CQDs and 480 nm and 580 nm for yellow CQDs. That is attributed to the large effective conjugation length and edge state from the conjugated pyrrole/pyrrolidone groups. When *m*-, *o*- or *p*-phenyldiamine was used as starting material, the absorption band of CQDs corresponding to the n → π* transition of conjugated C=N/C=O dramatically red-shifts from 355 nm to 426 and 512 for *m*-CQDs, *o*-CQDs and *p*-CQDs, respectively. The red-shift of this absorption band indicates the *p*-CQDs have smaller electronic band gap than others (Figure 3.15). The red-shift of emission maxima from λ = 435 nm, 535 nm and 604 nm for *m*-, *o*- and *p*-CQDs, repectively.

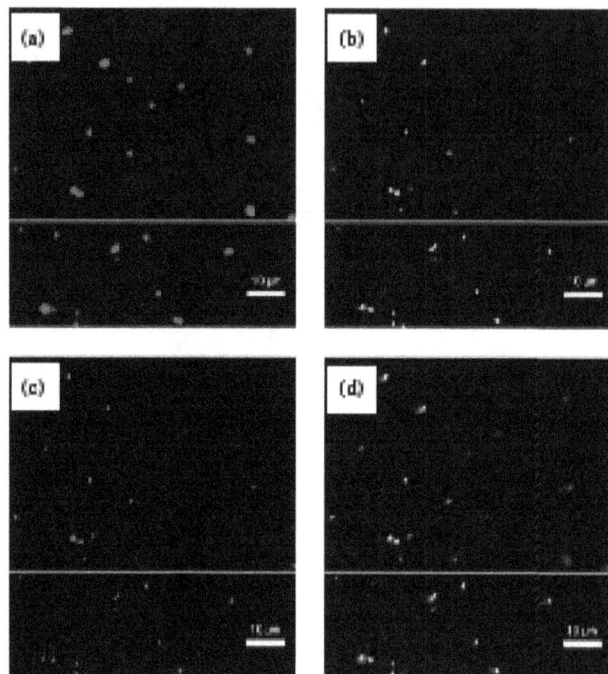

Figure 3.16 Single-particle FL emission images of the CQDs under excitations at (a) 360 nm, (b) 475 nm, (c) 535 nm, and (d) their overlaid image.[56] Reprinted with permission from Wiley.

The XPS results indicate that the pyridine N exists in the *o*- and *p*-CQDs besides the pyrrolic N in *m*-CQDs. These indicate that the construction of edge groups strongly affects the emission position.

Another unique optical property is that a single particle emits different color lights. Our group found that S, N co-doped CQDs prepared from solvothermal route can emit blue, green and red light with excitation at 360 nm, 460 nm and 560 nm from one CQD particle (Figure 3.7).[44] These emissions are nearly excitation-independent at 340–420 nm, 460–540 nm and 560–620 nm region for blue, green and red emissions. Recently, Lin's group observed this phenomenon, same particle can emit different color light under different excitation wavelengths (Figure 3.16).[56] That implied that the single particle could emit different color emissions.

3.4 Applications of CQDs

With their superior fluorescence, physiological stability, pH sensitivity, up-conversion PL property, fine biocompatibility and low toxicity, CQDs and their derivatives have offered substantial application perspectives. CQDs are generally non-toxic, possessing an overwhelming advantage over semiconductor quantum dots. For example, when 4,7,10-trioxa-1,13-tridecanediamine (TTDDA)-passivated CQDs were incubated at concentration levels below 500 mg mL^{-1} with HeLa cells for 24 h, the cell viability exceeded 90%.[57] Similarly, addition of CQDs to the culture medium containing human kidney cells did not induce significant cytotoxicity,[58] while no obvious organ damage was observed for mice treated with carboxylated CQDs.[59]

From the view of *in vivo* bioimaging, the light penetration depth through the tissue is a critical parameter for practical application. Generally, long wavelength light gives deeper penetration depth for the visible light. It is highly desired for preparing CQDs with red emission and green light excitation. Meanwhile, rapid development in synthesizing heteroatom-doped CQDs materials with tailorable functions as well as tunable PL further speeds up their applications. Until now, much effort has been centered on biological and photocatalytic applications.

3.4.1 *Bioimaging Application*

CQDs are promising candidates for biological applications, such as bioimaging, medical diagnostics, due to their low cytotoxicity, good chemical and photochemical stability and rich functional groups on the edge. Sun *et al.* first found the surface passivated CDs with PEG1500N are strongly emissive in the visible region, extending into the near-IR. Confocal microscopy images of *E. coli* ATCC 25922 labeled with the passivated CDs were obtained at different excitation wavelengths (Figure 3B).[60] It has been demonstrated that surface passivating agents of low cytotoxicity can be used safely at high concentrations for *in vivo* imaging. For example, PEGylated CQDs showed no noticeable toxic effects *in vivo* up to 28 days when mice were intravenously injected with 8–40 mg kg–1 (CQD/bodyweight) of the PEGylated CQDs into mice for toxicity evaluation.[61] No abnormalities were observed in harvested organs although the amounts of CQDs found in liver and spleen were higher than those found in other organs.[62] Our group showed the N-doped CQDs cultured with A549 cells for 24 h, there is no obvious change in the cell viability, which is over 90% even at 500 μg/mL. After incubation with N-GQDs (3 mg/mL) at 37°C for 1 h, the A549 cells under living conditions became brightly illuminated blue, green and red when imaged under microscope with different excitation wavelengths (405 nm, 488 nm and 555 nm), respectively (Figure 5). Similarly, S, N co-doped CQDs also exhibit low cytotoxicity on the HeLa cells and bright confocal imaging.[44,57] It was found that L929 cells retained 100% viability even in the presence of 2 mg mL^{-1} CQDs, and above 80% with 7 mg mL^{-1}. The CQD labeled L929 cells were further imaged *in vitro* utilizing confocal microscopy. L929 cells treated by CQDs showed bright green light under the excitation of 405 nm.[47] Se-doped CQDs also show good bioimaging behavior.[50] Besides the

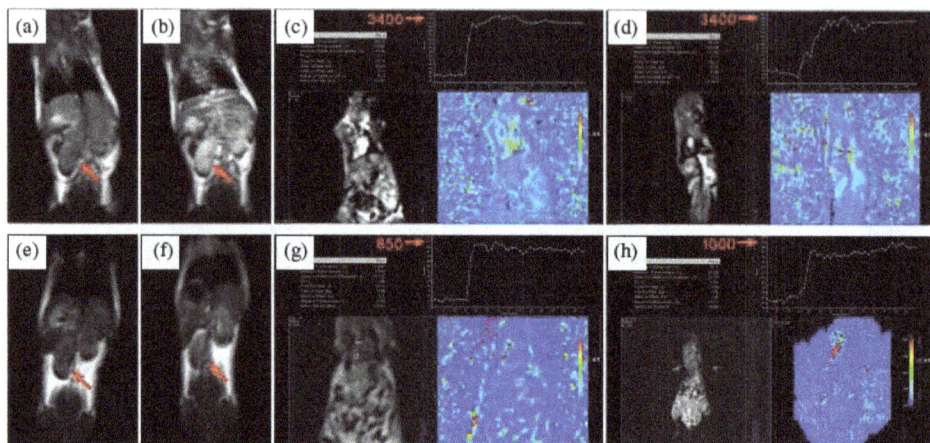

Figure 3.17 T1-weighted MR images of the mice injected with Gd-CQDs (a)–(d) and Gd-DTPA (Magnevists®) (e)–(h): (a), (e) plain scan, (b), (f) contrast-enhanced scan, (c), (g) cardiac perfusion scan, and (d), (h) aortic perfusion scan.[55] Reprinted with permission from RSC.

fluorescent imaging, the Gd-doped CQDs also exhibit good T1 contrast for dual fluorescent-MRI modes imaging capability.[53,54] Yi and coworkers compared the Gd-doped CQDs and commercial Gd-DTPA (Magnevist®). Both Gd-doped CQDs and Gd-DTPA entered the blood circulation of mice immediately after injection. The contrast-enhanced scan images (Figure 3.17(b)) revealed an obviously positive contrast enhancement of GCDs, as compared to plain scan images (Figure 3.17(a)). Similar phenomenon was observed for Gd-DTPA (Magnevists®) (Figures 3.17(e) and 3.17(f)). Perfusion imaging is a kind of functional MR imaging technique which can reveal the blood perfusion state and provides information towards hemodynamics. Thanks to the high r_1 relaxivity of GCDs, the peak intensity in cardiac and aortic perfusion scan reached 3400 in the mice injected with GCDs (Figures 3.17(c) and 3.17(d)), whereas the values generated by Gd-DTPA were only 650 and 1000 (Figures 3.17(g) and 3.17(h)), respectively. The significantly enhanced MR response by GCDs confirmed the great potential of GCDs in DCE perfusion MR imaging, especially in MR angiography.

3.4.2 Nanomedicine Based on CQDs

Fluorescent CQDs, being small nanoparticles, which can label the cell through endocytosis,[5,64] can be a perfect platform for constructing nanomedicine because of their low toxicity in the animals and optical properties.[65] Besides the bioimaging for diagnosis, CQDs can be used for photodynamic therapy, which is irradiated with a specific wavelength and can trigger the formation of singlet oxygen species that result in cell death. It has been confirmed that CQDs have high inhibition effect on MCF-7 and MDA-MB-231 cancer cells.[66] This phenomenon was attributed to

CQDs being able to generate more reactive oxygen species. Juzenas *et al.* also employed CQDs as photosensitizers in photodynamic therapy to destroy cancer cells.[67] CQDs functionalized with propionylethylenimine-co-ethylenimine-CQDs (PPEI-EI-CQDs) were prepared. Upon irradiation with UV light, these CQDs displayed substantial photodynamic effect in Du145 and PC3 cells. It was proposed that photo-induced generation of singlet oxygen and other reactive oxygen species and radicals are responsible for the observed photodynamic effect. By attaching a photosensitizer (chlorin e6) to CQDs, a synergistic photodynamic therapy platform was developed.[63] The CQDs-Ce6 conjugate is a good candidate with excellent imaging and remarkable photodynamic efficacy upon irradiation. CQDs–protoporphyrin (IX) sensitisier conjugate was also designed to exploit the large two-photon absorption cross section of CQDs and enable the indirect excitation of the sensitiser with 800 nm irradiation *via* FRET (Figure 3.18).[68]

Targeting function is an important property for diagnosis of tumor, delivery and release of nanomedicine. In particular, CQDs with fluorescence in the red light region would be the most desirable because background illumination from endogeneous fluorophores can be avoided during imaging.[70] It could result in many possibilities for conjugation with drug molecules in combination with targeting agents, expanding the drug choices for delivery. Hu and coworkers demonstrated that branch PEI-coated CQDs (bPEI-CQDs) displayed great potential in the application of gene delivery.[71] To make nanomedicine with targeting function, multiple steps are needed to attach the targeting molecules or peptide on the drug system. Sun *et al.* reported the fluorescent CQDs with self-targeting function are prepared from glucose and L-aspartic acid through one step thermolysis reaction.[69] The as-prepared CQDs-Asp show high selectivity on the brain tumor cells (C6 cells). Both *in vivo* and *in vitro* experiments demonstrated the CQDs-Asp can pass through the blood brain barrier and concentrate in the tumor site (Figure 3.19). This multiple function CQD will help to construct theranostic agent combined with fluorescent, targeting and therapeutic functions together and make the route to the personal customized nanomedicine.

The term "theranostics" was coined to define a proposed methodology that combined the modalities of the therapeutic and diagnostic.[72] Its goal is to develop specific and individualized therapeutic strategies towards personalized medicine, in light of the fact that acceptable efficacy of specific treatment could be achieved for only very few patients. By combining therapeutic and diagnostic capability into one single agent, a new protocol is anticipated to tailor a treatment based on the test results, thereby providing more specific and efficient systems for the curing of diseases. Such combination agents are materials capable of detecting and treating diseases in one single procedure. Therapeutic methods include chemo-, radio-, photothermal, photodynamic, gene and others. Diagnostic strategies are optical, computed tomography (CT), X-ray, photoacoustic and magnetic

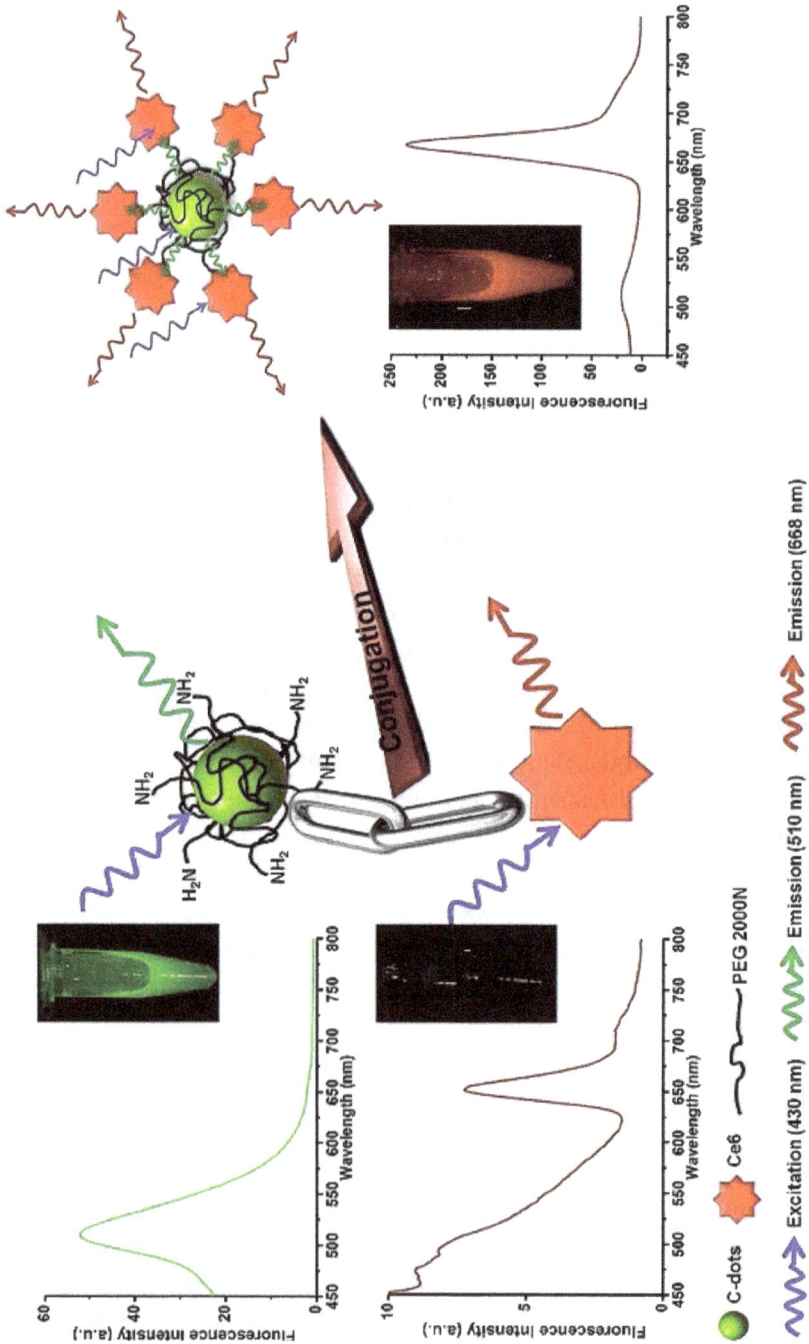

Figure 3.18   Forster resonance energy transfer (FRET) process between CQDs and Ce6.[63] Reprinted with permission from Wiley.

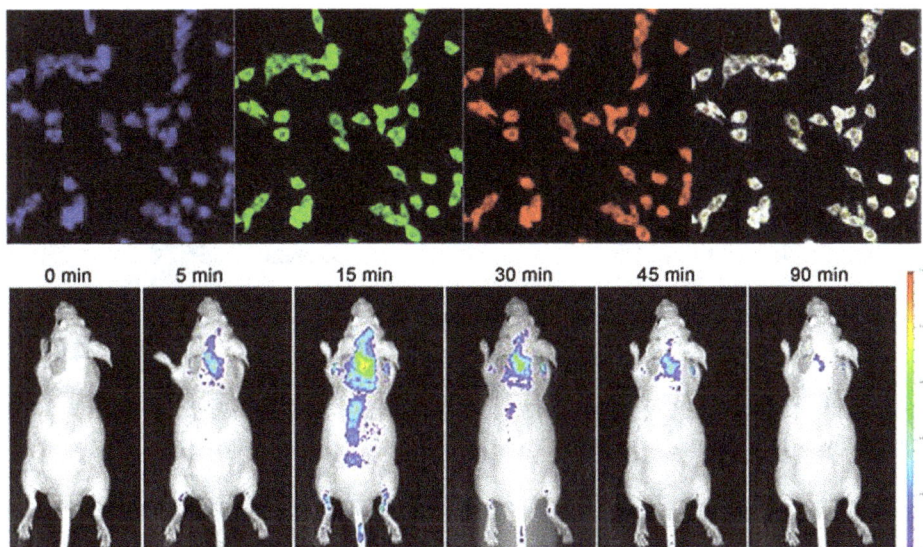

Figure 3.19 LSCM images of C6 cell lines pretreated with CD-Asp for 1 h at 37°C under excitation of 405 nm, 488 nm, and 555 nm (top). *In vivo* imaging of glioma-bearing mice after tail intravenous injection of CD-Asp (bottom). Whole body distribution of CD-Asp as a function of time after injection.[69] Reprinted with permission from ACS.

resonance (MR) imaging. CQDs, as a promising fluorescent imaging agent, could be employed to construct the theranostic agent.

Recently, our group successfully prepared a multifunctional theranostic agent (CQD-Oxa) by the conjugation of an anticancer agent (oxidized oxaliplatin, oxa(IV)–COOH) onto the surface of CQDs containing amine groups. CQD-Oxa successfully integrates the optical properties of the CQDs and the therapeutic performance of Oxa.[73] The *in vitro* results indicated that CQD-Oxa possesses good biocompatibility, bioimaging function, and anticancer effects. The pro-drug-conjugated CQDs were taken up by cancer cells through endocytosis and the drug was released upon the reduction of Oxa(IV)–COOH to oxaliplatin(II) because of the highly reducing environment in cancer cells. The *in vivo* results demonstrate that it is possible to follow the track or distribution of the drug by monitoring the fluorescence signal of CD-Oxa, which helps customize the injection time and dosage of the medicine (Figure 3.20). Karthik *et al.* demonstrated that a photoresponsive nanodrug delivery system was constructed using CQDs and a quinoline based phototrigger.[70] The strong fluorescent properties of CQDs have been explored for *in vitro* cellular imaging application, and the phototrigger ability of quinoline was exploited for efficient anticancer drug release using both one-photon and two-photon excitations. Folic acid is a common targeting molecule toward the tumor cells. Mewada and co-workers constructed the drug carrying and folic

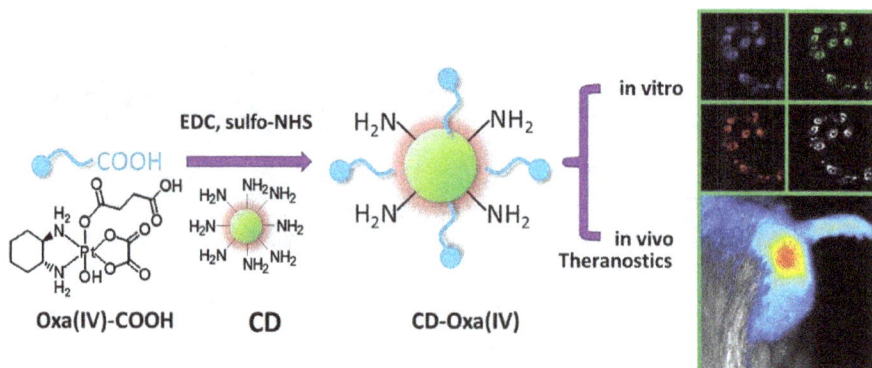

Figure 3.20   Synthetic scheme for CQD-Oxa and its applications in bioimaging and theranostics.[73] Reprinted with permission from Wiley.

acid-mediated delivering capacities of highly fluorescent swarming CQDs.[74] Folic acid on the CQD surface was used as a targeting molecule due to its high expression in most cancer cells. The drug loading capacity for an anti-cancer drug doxorubicin (DOX) was estimated to be B86% and the release of DOX from the DOX-loaded CQDs followed first order release kinetics at physiological pH — an ideal drug release profile DOX@CQDs showed a higher killing rate of cancer cells than free DOX and was found to be less toxic to normal cells due to FA mediated targeting.

3.4.3 *Photocatalytic Applications*

Photocatalysis is a typical artificial photosynthesis process. In nature, CO_2 and water are converted into glucose under sunlight with chlorophyll in chloroplast. H_2 produced by photocatalytic water splitting is one of the important renewable energy sources. On the other hand, photocatalytic degradation is also an effective way to the environmental therapy. The organic pollute can be degraded by active radical produced by photocatalyst under sunlight. Currently, most of the photocatalysts are wide band gap semiconductors, which only absorb the UV light. Because the CQD can absorb more broad light in the visible region, it can work as sensitizer, absorb more visible light and pass the photogenerated electron into the semiconductor to enhance its photocatalytic activity. Kang *et al.* loaded CQDs onto the anatase TiO_2 nanocrystals and observed the visible light H_2 production. Sun *et al.* loaded N and S, N co-doped CQD onto the P25 TiO_2 nanocrystals surface. The degradation of RhB reached ca. 30% and 60% for N-GQDs /TiO_2 and S, N-GQD/ TiO_2 composites, respectively. The apparent rate constant of S, N-GQD/TiO_2 is 0.01 which is 3 and 10 times higher than N-GQD/TiO_2 and TiO_2, respectively. The excellent degradation ability of S, N-GQD/TiO_2 is explained by S, N-GQD absorption properties in the visible region ($\lambda > 400$ nm) (qu nanoscale). S, N-CQDs/TiO_2 also showed enhanced H_2 production under $\lambda > 420$ nm visible light region.

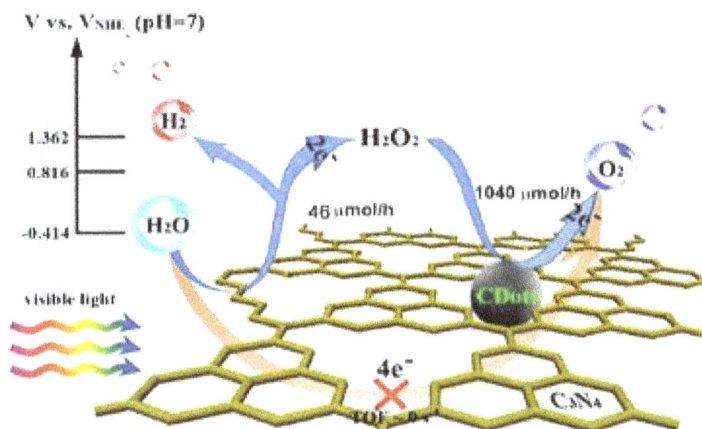

Figure 3.21 The proposed reaction mechanism for visible-light-driven water splitting by CQDs-C₃N₄.[75] Reprinted with permission from AAAS.

Besides CQDs working as a sensitizer, CQDs can also be good electron acceptors to promote the spatial separation of photo generated charge carriers. Recently, Kang *et al.* obtained the CQDs/g-C₃N₄ nanocomposites, which exhibit overall water splitting with the sunlight.[75] The quantum efficiencies of nanocomposites reached 16% for wavelength $\lambda = 420 \pm 20$ nm, 6.29% for $\lambda = 580 \pm 15$ nm, and 4.42% for $\lambda = 600 \pm 10$ nm, and they determined an overall solar energy conversion efficiency of 2.0%. More importantly, they proposed that a combination of CQDs and C3N4 could constitute a high-performance composite photocatalyst for water splitting *via* the stepwise two-electron/two-electron process (Figure 3.21): (i) $2H_2O \rightarrow H_2O_2 + H_2$; (ii) $2H_2O_2 \rightarrow 2H_2O + O_2$. The whole photocatalytic water splitting process is not through water oxidation process, which is one-step four-electron process.

3.5 Summary and Outlook

In this chapter, we have described the recent progress in the field of doped CQDs, focusing on their synthetic strategies, optical properties (absorption and PL), and applications in biomedicine, energy conversion issues.

Though several methods have been proposed towards the synthesis of high fluorescent CQDs, CQDs with high PL QY and IR or NIR emission are highly desired. It is critical to synthesize CQDs in a facile and green manner with designed structure and size for property studies and selected applications. Many studies have demonstrated the CQDs' versatility in biomedicine: (i) multimodal bioimaging for its flexibility in surface modification to combine other imaging agents for its high biocompatibility, (ii) multiple functions like targeting, therapy combining fluorescent CQDs together, (iii) delivery carrier for its various combinations with biomolecules or drugs *via* multi reaction and stimulus responses. It will arouse research interest in using CQDs in energy conversion due to their excellent electron acceptability and catalytic properties. The unique properties of CQDs entrust them with a wide field of research for bio and energy related applications.

References

1. Xu X., Ray R., Gu Y., Ploehn H. J., Gearheart L., Raker K. and Scrivens W. A. (2004). Electrophoretic analysis and purification of fluorescent single-walled carbon nanotube fragments, *J. Am. Chem. Soc.*, 126, 12736–12737.
2. Sun Y.-P., Zhou B., Lin Y., Wang W., Fernando K. A. S., Pathak P., Meziani M. J., Harruff B. A., Wang X., Wang H., Luo P. G., Yang H., Kose M. E., Chen B., Veca L. M. and Xie S.-Y. (2006). Quantum-sized carbon dots for bright and colorful photoluminescence, *J. Am. Chem. Soc.*, 128, 7756–7757.
3. Baker S. N. and Baker G. A. (2010). Luminescent carbon nanodots: Emergent nanolights, *Angew. Chem. Int. Ed.*, 49, 6726–6744.
4. Cao L., Meziani M. J., Sahu S. and Sun Y.-P. (2013). Photoluminescence properties of graphene *versus* other carbon nanomaterials, *Acc. Chem. Res.*, 46, 171–180.
5. Ding C., Zhu A. and Tian Y. (2014). Functional surface engineering of C-dots for fluorescent biosensing and *in vivo* bioimaging, *Acc. Chem. Res.*, 47, 20–30.
6. Zhang Z., Zhang J., Chen N. and Qu L. (2012). Graphene quantum dots: An emerging material for energy-related applications and beyond, *Energy Environ. Sci.*, 5, 8869–8890.
7. Hola K., Zhang Y., Wang Y., Giannelis E. P., Zboril R. and Rogach A. L. (2014). Carbon dots — Emerging light emitters for bioimaging, cancer therapy and optoelectronics, *Nano Today*, 9, 590–603.
8. Tang L., Ji R., Cao X., Lin J., Jiang H., Li X., Teng K. S., Luk C. M., Zeng S., Hao J. and Lau S. P. (2012). Deep ultraviolet photoluminescence of water-soluble self-passivated graphene quantum dots, *ACS Nano*, 6, 5102–5110.
9. Resch-Genger U., Grabolle M., Cavaliere-Jaricot S., Nitschke R. and Nann T. (2008). Quantum dots *versus* organic dyes as fluorescent labels, *Nat. Meth.*, 5, 763–775.
10. Alivisatos A. P., Gu W. and Larabell C. (2005). Quantum dots as cellular probes, *Annu. Rev. Biomed. Eng.*, 7, 55–76.
11. Zheng X. T., Ananthanarayanan A., Luo K. Q. and Chen P. (2015). Glowing graphene quantum dots and carbon dots: Properties, syntheses, and biological applications, *Small*, 11, 1620–1636.
12. Li X., Rui M., Song J., Shen Z. and Zeng H. (2015). Carbon and graphene quantum dots for optoelectronic and energy devices: A review, *Adv. Funct. Mater.*, 25, 4929–4947.
13. Shen J., Zhu Y., Yang X. and Li C. (2012). Graphene quantum dots: Emergent nanolights for bioimaging, sensors, catalysis and photovoltaic devices, *Chem. Commun.*, 48, 3686–3699.

14. Zheng L., Chi Y., Dong Y., Lin J. and Wang B. (2009). Electrochemiluminescence of water-soluble carbon nanocrystals released electrochemically from graphite, *J. Am. Chem. Soc.*, 131, 4564–4565.
15. Li H., He X., Kang Z., Huang H., Liu Y., Liu J., Lian S., Tsang C. H. A., Yang X. and Lee S.-T. (2010). Water-soluble fluorescent carbon quantum dots and photocatalyst design, *Angew. Chem., Int. Ed. Engl.*, 49, 4430–4434.
16. Lin L. and Zhang S. (2012). Creating high yield water soluble luminescent graphene quantum dots *via* exfoliating and disintegrating carbon nanotubes and graphite flakes, *Chem. Commun.*, 48, 10177–10179.
17. Zhou J., Booker C., Li R., Zhou X., Sham T.-K., Sun X. and Ding Z. (2007). An electrochemical avenue to blue luminescent nanocrystals from multiwalled carbon nanotubes (MWCNTs), *J. Am. Chem. Soc.*, 129, 744–745.
18. Peng J., Gao W., Gupta B. K., Liu Z., Romero-Aburto R., Ge L., Song L., Alemany L. B., Zhan X., Gao G., Vithayathil S. A., Kaipparettu B. A., Marti A. A., Hayashi T., Zhu J.-J. and Ajayan P. M. (2012). Graphene quantum dots derived from carbon fibers, *Nano Lett.*, 12, 844–849.
19. Qiao Z.-A., Wang Y., Gao Y., Li H., Dai T., Liu Y. and Huo Q. (2010). Commercially activated carbon as the source for producing multicolor photoluminescent carbon dots by chemical oxidation, *Chem. Commun.*, 46, 8812–8814.
20. Ye R., Xiang C., Lin J., Peng Z., Huang K., Yan Z., Cook N. P., Samuel E. L. G., Hwang C.-C., Ruan G., Ceriotti G., Raji A.-R. O., Martí A. A. and Tour J. M. (2013). Coal as an abundant source of graphene quantum dots, *Nat. Commun.*, 4,
21. Hu C., Yu C., Li M., Wang X., Yang J., Zhao Z., Eychmüller A., Sun Y. P. and Qiu J. (2014). Chemically tailoring coal to fluorescent carbon dots with tuned size and their capacity for Cu(II) detection, *Small*, 10, 4926–4933.
22. Dong Y., Shao J., Chen C., Li H., Wang R., Chi Y., Lin X. and Chen G. (2012). Blue luminescent graphene quantum dots and graphene oxide prepared by tuning the carbonization degree of citric acid, *Carbon*, 50, 4738–4743.
23. Wang F., Xie Z., Zhang H., Liu C.-y. and Zhang Y.-g. (2011). Highly luminescent organosilane-functionalized carbon dots, *Adv. Funct. Mater.*, 21, 1027–1031.
24. Zhu H., Wang X., Li Y., Wang Z., Yang F. and Yang X. (2009). Microwave synthesis of fluorescent carbon nanoparticles with electrochemiluminescence properties, *Chem. Commun.*, 5118–5120.
25. Li H., He X., Liu Y., Huang H., Lian S., Lee S.-T. and Kang Z. (2011). One-step ultrasonic synthesis of water-soluble carbon nanoparticles with excellent photoluminescent properties, *Carbon*, 49, 605–609.
26. Jia X., Li J. and Wang E. (2012). One-pot green synthesis of optically pH-sensitive carbon dots with upconversion luminescence, *Nanoscale*, 4, 5572–5575.
27. Sahu S., Behera B., Maiti T. K. and Mohapatra S. (2012). Simple one-step synthesis of highly luminescent carbon dots from orange juice: Application as excellent bio-imaging agents, *Chem. Commun.*, 48, 8835–8837.
28. Wang J., Wang C.-F. and Chen S. (2012). Amphiphilic egg-derived carbon dots: Rapid plasma fabrication, pyrolysis process, and multicolor printing patterns, *Angew. Chem., Int. Ed.*, 51, 9297–9301.

29. Yang Y., Cui J., Zheng M., Hu C., Tan S., Xiao Y., Yang Q. and Liu Y. (2012). One-step synthesis of amino-functionalized fluorescent carbon nanoparticles by hydrothermal carbonization of chitosan, *Chem. Commun.*, 48, 380–382.

30. Qu D., Zheng M., Zhang L., Zhao H., Xie Z., Jing X., Haddad R. E., Fan H. and Sun Z. (2014). Formation mechanism and optimization of highly luminescent N-doped graphene quantum dots, *Sci. Rep.*, 4, 5294.

31. Bourlinos A. B., Stassinopoulos A., Anglos D., Zboril R., Karakassides M. and Giannelis E. P. (2008). Surface functionalized carbogenic quantum dots, *Small*, 4, 455–458.

32. Bourlinos A. B., Zbořil R., Petr J., Bakandritsos A., Krysmann M. and Giannelis E. P. (2012). Luminescent surface quaternized carbon dots, *Chem. Mater.*, 24, 6–8.

33. Bhunia S. K., Saha A., Maity A. R., Ray S. C. and Jana N. R. (2013). Carbon nanoparticle-based fluorescent bioimaging probes, *Sci. Rep.*, 3, 1473.

34. Zheng M., Xie Z., Qu D., Li D., Du P., Jing X. and Sun Z. (2013). On–off–on fluorescent carbon dot nanosensor for recognition of Chromium(VI) and ascorbic acid based on the inner filter effect, *ACS Appl. Mater. Interfaces*, 5, 13242–13247.

35. Zhu S., Meng Q., Wang L., Zhang J., Song Y., Jin H., Zhang K., Sun H., Wang H. and Yang B. (2013). Highly photoluminescent carbon dots for multicolor patterning, sensors, and bioimaging, *Angew. Chem., Int. Ed.*, 52, 3953–3957.

36. Qu D., Zheng M., Li J., Xie Z. and Sun Z. (2015). Tailoring color emissions from N-doped graphene quantum dots for bioimaging applications, *Light Sci Appl*, 4, e364.

37. Jiang K., Sun S., Zhang L., Lu Y., Wu A., Cai C. and Lin H. (2015). Red, green, and blue luminescence by carbon dots: Full-color emission tuning and multicolor cellular imaging, *Angew. Chem., Int. Ed. Engl.*, 54, 5360–5363.

38. Li S., Li Y., Cao J., Zhu J., Fan L. and Li X. (2014). Sulfur-doped graphene quantum dots as a novel fluorescent probe for highly selective and sensitive detection of Fe3+, *Anal. Chem.*, 86, 10201–10207.

39. Yang S., Sun J., Li X., Zhou W., Wang Z., He P., Ding G., Xie X., Kang Z. and Jiang M. (2014). Large-scale fabrication of heavy doped carbon quantum dots with tunable-photoluminescence and sensitive fluorescence detection, *J. Mater. Chem. A*, 2, 8660–8667.

40. Xu Q., Pu P., Zhao J., Dong C., Gao C., Chen Y., Chen J., Liu Y. and Zhou H. (2015). Preparation of highly photoluminescent sulfur-doped carbon dots for Fe(III) detection, *J. Mater. Chem. A*, 3, 542–546.

41. Dong Y., Pang H., Yang H. B., Guo C., Shao J., Chi Y., Li C. M. and Yu T. (2013). Carbon-based dots co-doped with nitrogen and sulfur for high quantum yield and excitation-independent emission, *Angew. Chem., Int. Ed. Engl.*, 52, 7800–7804.

42. Qu D., Zheng M., Du P., Zhou Y., Zhang L., Li D., Tan H., Zhao Z., Xie Z. and Sun Z. (2013). Highly luminescent S, N co-doped graphene quantum dots with broad visible absorption bands for visible light photocatalysts, *Nanoscale*, 5, 12272–12277.

43. Ding H., Wei J. S. and Xiong H. M. (2014). Nitrogen and sulfur co-doped carbon dots with strong blue luminescence, *Nanoscale*, 6, 13817–13823.

44. Qu D., Sun Z., Zheng M., Li J., Zhang Y., Zhang G., Zhao H., Liu X. and Xie Z. (2015). Three Colors Emission from S,N co-doped graphene quantum dots for visible light H2 production and bioimaging, *Adv. Opt. Mater.*, 3, 360–367.

45. Zhou J., Shan X., Ma J., Gu Y., Qian Z., Chen J. and Feng H. (2014). Facile synthesis of P-doped carbon quantum dots with highly efficient photoluminescence, *RSC Adv.*, 4, 5465–5468.

46. Xu Z.-Q., Yang L.-Y., Fan X.-Y., Jin J.-C., Mei J., Peng W., Jiang F.-L., Xiao Q. and Liu Y. (2014). Low temperature synthesis of highly stable phosphate functionalized two color carbon nanodots and their application in cell imaging, *Carbon*, 66, 351–360.

47. Wang W., Li Y., Cheng L., Cao Z. and Liu W. (2014). Water-soluble and phosphorus-containing carbon dots with strong green fluorescence for cell labeling, *J. Mater. Chem. B*, 2, 46–48.

48. Qian Z., Shan X., Chai L., Ma J., Chen J. and Feng H. (2014). Si-doped carbon quantum dots: A facile and general preparation strategy, bioimaging application, and multifunctional sensor, *ACS Appl. Mater. Interfaces*, 6, 6797–6805.

49. Li X., Lau S. P., Tang L., Ji R. and Yang P. (2013). Multicolour light emission from chlorine-doped graphene quantum dots, *J. Mater. Chem. C*, 1, 7308–7313.

50. Yang S., Sun J., He P., Deng X., Wang Z., Hu C., Ding G. and Xie X. (2015). Selenium doped graphene quantum dots as an ultrasensitive redox fluorescent switch, *Chem. Mater.*, 27, 2004–2011.

51. Fan Z., Li Y., Li X., Fan L., Zhou S., Fang D. and Yang S. (2014). Surrounding media sensitive photoluminescence of boron-doped graphene quantum dots for highly fluorescent dyed crystals, chemical sensing and bioimaging, *Carbon*, 70, 149–156.

52. Shen P. and Xia Y. (2014). Synthesis-modification integration: One-step fabrication of boronic acid functionalized carbon dots for fluorescent blood sugar sensing, *Anal. Chem.*, 86, 5323–5329.

53. Bourlinos A. B., Bakandritsos A., Kouloumpis A., Gournis D., Krysmann M., Giannelis E. P., Polakova K., Safarova K., Hola K. and Zboril R. (2012). Gd(III)-doped carbon dots as a dual fluorescent-MRI probe, *J. Mater. Chem.*, 22, 23327–23330.

54. Gong N., Wang H., Li S., Deng Y., Chen X. a., Ye L. and Gu W. (2014). Microwave-assisted polyol synthesis of gadolinium-doped green luminescent carbon dots as a bimodal nanoprobe, *Langmuir*, 30, 10933–10939.

55. Pan Y., Yang J., Fang Y., Zheng J., Song R. and Yi C. (2017). One-pot synthesis of gadolinium-doped carbon quantum dots for high-performance multimodal bioimaging, *J. Mater. Chem. B*, 5, 92–101.

56. Pan L., Sun S., Zhang A., Jiang K., Zhang L., Dong C., Huang Q., Wu A. and Lin H. (2015). Truly fluorescent excitation-dependent carbon dots and their applications in multicolor cellular imaging and multidimensional sensing, *Adv. Mater.*, 27, 7782–7787.

57. Sun D., Ban R., Zhang P.-H., Wu G.-H., Zhang J.-R. and Zhu J.-J. (2013). Hair fiber as a precursor for synthesizing of sulfur- and nitrogen-co-doped carbon dots with tunable luminescence properties, *Carbon*, 64, 424–434.

58. Zhao Q.-L., Zhang Z.-L., Huang B.-H., Peng J., Zhang M. and Pang D.-W. (2008). Facile preparation of low cytotoxicity fluorescent carbon nanocrystals by electrooxidation of graphite, *Chem. Commun.*, 5116–5118.

59. Nurunnabi M., Khatun Z., Huh K. M., Park S. Y., Lee D. Y., Cho K. J. and Lee Y.-k. (2013). *In vivo* biodistribution and toxicology of carboxylated graphene quantum dots, *ACS Nano*, 7, 6858–6867.

60. Cao L., Wang X., Meziani M. J., Lu F., Wang H., Luo P. G., Lin Y., Harruff B. A., Veca L. M., Murray D., Xie S.-Y. and Sun Y.-P. (2007). Carbon dots for multiphoton bio-imaging, *J. Am. Chem. Soc.*, 129, 11318–11319.

61. Wang Y., Anilkumar P., Cao L., Liu J.-H., Luo P. G., Tackett K. N., Sahu S., Wang P., Wang X. and Sun Y.-P. (2011). Carbon dots of different composition and surface functionalization: cytotoxicity issues relevant to fluorescence cell imaging, *Exp. Biol. Med.*, 236, 1231–1238.

62. Yang S.-T., Wang X., Wang H., Lu F., Luo P. G., Cao L., Meziani M. J., Liu J.-H., Liu Y., Chen M., Huang Y. and Sun Y.-P. (2009). Carbon dots as nontoxic and high-performance fluorescence imaging agents, *J. Phys. Chem. C*, 113, 18110–18114.

63. Huang P., Lin J., Wang X., Wang Z., Zhang C., He M., Wang K., Chen F., Li Z., Shen G., Cui D. and Chen X. (2012). Light-triggered theranostics based on photosensitizer-conjugated carbon dots for simultaneous enhanced-fluorescence imaging and photodynamic therapy, *Adv. Mater.*, 24, 5104–5110.

64. Li H., Kang Z., Liu Y. and Lee S.-T. (2012). Carbon nanodots: Synthesis, properties and applications, *J. Mater. Chem.*, 22, 24230–24253.

65. Tao H., Yang K., Ma Z., Wan J., Zhang Y., Kang Z. and Liu Z. (2012). *In vivo* nir fluorescence imaging, biodistribution, and toxicology of photoluminescent carbon dots produced from carbon nanotubes and graphite, *Small*, 8, 281–290.

66. Hsu P.-C., Chen P.-C., Ou C.-M., Chang H.-Y. and Chang H.-T. (2013). Extremely high inhibition activity of photoluminescent carbon nanodots toward cancer cells, *J. Mater. Chem. B*, 1, 1774–1781.

67. Juzenas P., Kleinauskas A., George Luo P. and Sun Y.-P. (2013). Photoactivatable carbon nanodots for cancer therapy, *Appl. Phys. Lett.*, 103, 063701.

68. Fowley C., Nomikou N., McHale A. P., McCaughan B. and Callan J. F. (2013). Extending the tissue penetration capability of conventional photosensitisers: A carbon quantum dot-protoporphyrin IX conjugate for use in two-photon excited photodynamic therapy, *Chem. Commun.*, 49, 8934–8936.

69. Zheng M., Ruan S., Liu S., Sun T., Qu D., Zhao H., Xie Z., Gao H., Jing X. and Sun Z. (2015). Self-targeting fluorescent carbon dots for diagnosis of brain cancer cells, *ACS Nano*, 9, 11455–11461.

70. Karthik S., Saha B., Ghosh S. K. and Pradeep Singh N. D. (2013). Photoresponsive quinoline tethered fluorescent carbon dots for regulated anticancer drug delivery, *Chem. Commun.*, 49, 10471–10473.

71. Hu L., Sun Y., Li S., Wang X., Hu K., Wang L., Liang X.-j. and Wu Y. (2014). Multifunctional carbon dots with high quantum yield for imaging and gene delivery, *Carbon*, 67, 508–513.

72. Funkhouser J. (2002). Reinventing pharma: The theranostic revolution, *Curr. Drug Discov.*, 2, 17–19.

73. Zheng M., Liu S., Li J., Qu D., Zhao H., Guan X., Hu X., Xie Z., Jing X. and Sun Z. (2014). Integrating oxaliplatin with highly luminescent carbon dots: An unprecedented theranostic agent for personalized medicine, *Adv. Mater.*, 26, 3554–3560.

74. Mewada A., Pandey S., Thakur M., Jadhav D. and Sharon M. (2014). Swarming carbon dots for folic acid mediated delivery of doxorubicin and biological imaging, *J. Mater. Chem. B*, 2, 698–705.

75. Liu J., Liu Y., Liu N., Han Y., Zhang X., Huang H., Lifshitz Y., Lee S.-T., Zhong J. and Kang Z. (2015). Metal-free efficient photocatalyst for stable visible water splitting *via* a two-electron pathway, *Science*, 347, 970–974.

Index

www.ingramcontent.com/pod-product-compliance
Lightning Source LLC
Chambersburg PA
CBHW081520190326

41458CB00015B/5414